AF706841

Asia's Entrepreneurs

This book is a collection of technology start-up cases in Asia, told in a narrative form, to give readers an insider view to how innovators and technopreneurs view entrepreneurial opportunities from the use of technology to how the technopreneurs raise funding to support their vision, and the subsequent relationship of the technopreneurs and their investors. The book illustrates some of the cases using the theory of effectuation.

The book can be used by practitioners and by educators for developing a deep understanding on the issues of raising capital for the purpose of growing the venture, and the trade-offs of capital from the different groups of investors and their financial terms. It is useful to young and mid-career professionals looking at starting a technology venture in Asia.

Virginia Cha is the Chief of Research Practice at the Institute of Systems Science, National University of Singapore (NUS). She is an active researcher, educator, mentor, and angel investor in Singapore's entrepreneur ecosystem, with 32 years of executive management experience in technology companies. Previous roles include founder and CEO of venture-funded, hi-tech companies in Singapore and China with successful exits on NASDAQ and HKSE. In addition, she is an active mentor and angel investor to early stage start-ups, with focus in technology ventures at INSEAD, NUS, and Singapore Management University. She is a member of the Global Agenda Council on Fostering Entrepreneurship 2012–14, World Economic Forum.

Jennifer Lien is a Research Assistant at the Institute of Systems Science at the National University of Singapore. She has previously held positions in private education and in municipal affairs, and is an active young researcher and community builder.

Asia's Entrepreneurs

Dilemmas, risks, and opportunities

Virginia Cha and Jennifer Lien

Routledge
Taylor & Francis Group
LONDON AND NEW YORK

First published 2013
by Routledge

2 Park Square, Milton Park,
Abingdon, Oxfordshire OX14 4RN

Simultaneously published in the USA and Canada
by Routledge
711 Third Avenue, New York, NY 10017

First issued in paperback 2014

Routledge is an imprint of the Taylor and Francis Group, an informa business

British Library Cataloguing in Publication Data

A catalogue record for this book is available from the British Library

Library of Congress Cataloging-in-Publication Data
Cha, Virginia.
Asia's entrepreneurs: dilemmas, risks, and opportunities/by Virginia Cha and Jennifer Lien.
p. cm.
1. Entrepreneurship–Asia. 2. New business enterprises–Asia. 3. New business enterprises–Asia–Finance. 4. High technology industries–Asia. I. Lien, Jennifer, 1974- II. Virginia Cha.
HB615.C517 2013
338'.04095–dc23
2012038590

ISBN 978-0-415-52276-2 (hbk)
ISBN 978-1-138-91037-9 (pbk)
ISBN 978-0-203-55685-6 (ebk)

Typeset in Times New Roman
by OKS Prepress Services, Chennai, India

Note from the authors

This book is a compilation of selected case histories of Singapore's technopreneurs, told in a narrative form. The authors captured the recollections of each case protagonist to the best of our abilities. The reader should be alerted that all recollections are subject to errors. Therefore, some factual inaccuracies are expected.

Contents

Foreword

I should start by commending Virginia for penning some of the key events in Singapore's journey towards building an entrepreneurial ecosystem, beginning with her own entrepreneurial journey.

The mid 1990s were the heady days when Singapore was mounting a multipronged approach towards reshaping and re-engineering its economy. There was a spirited and intense effort in nurturing and growing a life science industry in Singapore under the leadership of Singapore's Economic Development Board. There was a second effort to grow local high technology start-ups by a team headed by the then Deputy Prime Minister and now the President of Singapore, Dr Tony Tan, ably assisted by Mr Teo Ming Kian. This team's effort was most concerned with capturing economic value from the investments into the different institutes under the then National Science and Technology Board (NSTB), the predecessor to A*STAR (Agency for Science Technology and Research). This programme was named Technopreneurship, to indicate the efforts to persuade engineers and scientist to consider pursuing an entrepreneurship career.

Technopreneurship was positioned as a catalyst to enable the flow of intellectual property created at the NSTB institutes into the market place. A number of support structures were established including Venture TDF as a direct investor and Technopreneurship Investment Fund (TIF) as a fund of funds. A number of institutes responded to this new direction including Kent Ridge Digital Labs. Kent Ridge Digital Labs was an amalgamation of the research division of the Institute of Systems Science and the Information Technology Institute. The former group had distinguished itself at creating state of the art technologies while the latter had developed innovative applications deployed in Singapore. There was a general air of great expectations from the union of these two institutions that the two distinct set of competencies housed together would result in exciting innovations of global applications. The resulting entity did not disappoint its creators. They managed to spin off around 20 technology-based companies, which attracted substantial investments from venture capitalists and corporate investors.

There was so much confidence and optimism in the successful creation of technology-based start-ups that there was even a plan to spin-off a 'high tech company' generator using the intellectual property of one of the institutes, possibly aided by management talent from one of the large government-linked companies. The plan was blessed by those who need to bless such plans and investors were sought to fund this entity. A management team for the new entity was identified, as was key talent who would move from the institute to the new entity.

The 2000 dotcom bust dampened the spirits of the technopreneurship movement. Highest level reviews led to an eventual hiatus of this movement. Only historians can tell whether that was the best time to have pushed on, given that talent was available at ever low prices and a number of other resources required for the creation of high tech start-ups was available in plenty and at lower than normal costs.

Technopreneurship resurfaced in 2006 under the aegis of National Research Foundation (NRF), once again under the leadership of Dr Tony Tan and Mr Teo Ming Kian. NRF's primary role was to catalyze new industries and develop innovative solutions to Singapore's national challenges. NRF aspired to create a vibrant R&D-based innovation system for a knowledge-based Singapore economy. They approached this challenge from both top-down and bottom-up programs. The top-down approach resulted in establishing strategic research programs and national innovation challenges. The bottom-up approach designed and implemented a Campus for Research Excellence and Technological Enterprise (CREATE), NRF Fellowship, Competitive Research Programme, Research Centers of Excellence and defined a National Framework for Innovation and Enterprise (NFIE).

NRF established a $350 million fund under the NFIE to promote academic entrepreneurship that supported a number of programs. These include university innovation funds whose primary objective was to create an innovation and entrepreneurship buzz on the campuses of the institutions of higher learning (IHLs). This was augmented with proof of concept funding, a technology incubator scheme and early-stage venture capital funding. Proof of concept funding was established to move intellectual property from the laboratories of the IHLs into the market place. The technology incubator scheme provided significant government investment leverage to private investors who were willing to establish incubators for creating high technology companies. Early stage venture capital fund provided a one to one matching to venture capital funds that were willing to invest in translating interesting technologies into successful businesses.

Virginia's book is set against this backdrop of a critical period in the high tech entrepreneurship history of Singapore.

Singapore had a first generation of technology entrepreneurs in Michael Mun of Aztech, Sim Wong Hoo, Ng Kai Wa and Chay Kwong Soon of Creative Technologies, Ngiam brothers of IPC, Ron Sim of Osim, and Henn Tan of Trek 2000, later joined by Olivia Lum of Hyflux. Ong Peng Tsin is another entrepreneur who succeeded in ventures such as Match.com, Interwoven, both created successfully for the US markets. He went on to create Encentuate out of Singapore which was eventually bought out by IBM. These are well-known names. Virginia's book introduces another generation of entrepreneurs, primarily in the ICT space. I have been fortunate in knowing most of those mentioned in this book personally and have seen them succeed in their ventures to different degrees. I like the way Virginia has organized her book – Part I focused on getting the first partner right, Part II on managing dilemmas in order to keep afloat and Part III on the coaches and the incubation programmers.

I should also mention a couple serial entrepreneurs to complete the list – Dr Ting Choon Meng and Eddie Chau. Dr Ting has created four medical device companies and has perhaps created the gold standard in the management of hypertension. He is also an expert in knee-related problems. His latest company, Healthstats International, appears to be heading towards creating a global impact. Eddie created e-cop and went on to create Brandtology.

I believe Virginia's book is a must-read for those interested in knowing a slice of Singapore's recent entrepreneurial history. I wish her every success with this book and look forward to reading her future creations.

Arcot Desai Narasimhalu
Director
Institute of Innovation and Entrepreneurship
Singapore Management University

Preface

I want to start off this book by stating that I am not a particularly successful technology entrepreneur by the conventional yardstick. I did not create a company with a billion (name your favourite currency: baht/dollar/ringgit/rupiah/yen/yuan) IPO, nor did I get fabulously wealthy from any of my venture exits. This should be immediately obvious since filthy-rich entrepreneurs would not sit still to write a book until post-retirement. My journey, in many respects, reflects that of a typical serial entrepreneur: with starts, pivots, missteps, exits, and even one un-ceremonial boot-out. I started this journey in Singapore, in 1995, when I gave up my corporate executive position at Unisys Corporation to move here from the USA. I was recruited by the Institute of Systems Science (ISS)[1] to commercialize IT technologies developed in the research lab, funded by the then-National Science Technology Board (NSTB).[2] Never one to stay behind when there are new mountains to climb, I left the ISS to create the first venture-funded spin-off from a national-funded lab technology in 1997. I named this new venture Star + Globe Technologies as a pun on the old style Chinese name for Singapore (星加坡), when pronounced in Cantonese dialect. I thought this would be an appropriate name since the venture used the multilingual information processing technologies from the lab with the aspiration of creating a global village where language barriers would be removed by use of our applications. This was in the heady days of the dawn of business information migrating to the internet. We had an exhilarating ride with Star + Globe, cumulating with her being reverse-merged with a NASDAQ-listed company at the height of the 1999 dotcom bubble. After my exit (Chapter 1) from the company, I was recruited to China by a private equity investment company and subsequently started more companies.

In China, I met and befriended many entrepreneurs, including the now very famous Carson Block from Muddy Waters Research of Chinese company fraud exposé fame, and Marc Van Der Chijs, co-founder of Tudou and now founder of UnitedStyles. I will write the China entrepreneur stories in another book, with the tentative title *The Bud Fox*[3] *Society* already picked out giving a hint on the theme of that future book. The lessons learned from the China days are relevant to my personal journey and subsequent return to Singapore in 2007, exactly a decade after my first start-up. I decided to write my first book on Singapore's technopreneurs – the stories herein draw a fascinating picture of the many entrepreneur stories from the city-state, my adopted home, and her aspiration to create an entrepreneur ecosystem.

After more than 10 years of creating successive new ventures and constantly being in the community of entrepreneurs, I became very curious about entrepreneurship, in the Asian and especially Singapore context. Hence, I enrolled in the PhD program at the National University of Singapore (NUS) and started to embark on an intellectual and systematic understanding of technology entrepreneurship. Fast forward to 2012, I re-emerged back into the entrepreneur ecosystem *jianghu*,[4] armed with a PhD degree and formal theories on entrepreneurship, and began to run again with technology entrepreneurs – this time as a mentor, advisor, and angel investor.

This is a book unique in many respects: it is a mash-up of my personal interpretations of entrepreneurship as a participant, and moderated by my formal training in theoretical frameworks, research methodology, and philosophy of science. This book is about technology entrepreneurship over the past 30 years in Singapore, an under-published area considering the level of entrepreneur activity and the active but underreported transformation over this period. There is a dearth of information on failed start-up cases, perhaps as an Asian 'face-saving' mechanism. So, this book is not only about successful exits, but also about missteps and missed opportunities in Asia. Hopefully, the lessons drawn from both the successful and the not-so-successful ventures will be insightful to students, entrepreneurs, and all others interested in learning about what it takes to create a technology venture in the Asian context.

In the 1980s and 1990s, technology start-ups in Singapore were few and far between. Creative Technologies remained the lone leader from Singapore through those decades. Globally, the more common technology-based start-ups were spin-outs from large multinational technology companies. The more famous high tech spin-outs were from Palo Alto Research Centre, a Xerox company referred to as PARC. From its inception in 1970, it has invented some of today's mainstays, such as the PDF file format and the laser printer. Even Apple's now famous point and click user interface was a technology invented at PARC. The spin-out literature[5] characterizes the phenomenon of new firms spawning from former employees either as a Fairchild or a Xerox model. With the Fairchild model, spin-outs occurred mostly with effective employee learning and positive external environment conducive to new venture creation, which resulted in the concentration of technology clusters in the Silicon Valley and Boston's Route 128 in the USA. By contrast, the Xerox model attributed the spin-outs to employee disagreement on the value of the innovation. Both models assumed that employees would have access to new ideas, concepts, and innovations during employment, and have the opportunity for the new venture creation – either from employee learning or from incumbent under-exploitation – prior to venture spin-outs.

Prior to the widespread availability of information from the internet, exposure to new innovations and technology inventions was uncommon for many nascent entrepreneurs in Asia. In Singapore, technology-enabled opportunity as a pull factor was seen as playing a less important role in new venture creation. Hence, many started as imitative or incremental innovation firms since entrepreneurs started new firms based on spotting a gap in the line-up of technology offerings or in the market segmentation during their employment in the larger company. I will chronicle many of these spin-out start-ups, viewed from the lens of the technopreneurs who left a comfortable corporate position to brave entrepreneurship. One common theme for this earlier batch (pre-2000) of technopreneurs is constraints from the lack of a vibrant venture funding ecosystem. For many of them, this was a major barrier for starting innovative technology firms, which subsequently limited their growth. For the few who did manage to secure venture funding, they found themselves having to manage investor expectations, an activity that was not given sufficient import by them.

Fast forward to today where many technology incubation and funding programs are in place, newer start-ups are now competing on innovations in products/services or business models. I have had the good fortune for being a participant in a deep-technology start-up from the old days, and was able to interview many spin-out entrepreneurs from the pre-2000 period from their perspective. Today, I am an active participant in the new generation start-up scene as a mentor, advisor, and angel investor to early stage start-ups at INSEAD, NUS, and Singapore Management University (SMU). From the cases I have interviewed and from insights gained from active mentoring activities, I was able to reflect upon my own entrepreneurship journey and those of my contemporaries from the mid-1990s and compare them against the new generation of start-ups. This reflection allowed me to appreciate the vast transformation of Singapore's entrepreneurship scene in past three decades.

For the stories and lessons captured in this book, I drew on the lessons from over 99 entrepreneurs on their entrepreneurial journey. I selected nine cases from this large case load and created seven narrative cases and two teaching cases[6] to show case the dilemmas, risks, and opportunities experienced by them. This book is divided into four parts. Part I is about technopreneurs and their investors – the missed expectations so common for early-stage technology start-ups and the untimely interventions by investors. Part II describes how some of the technopreneurs from the same era pre-2000 managed dilemmas effectively and persevered to find a profitable exit. Part III is about how Singapore's entrepreneurial ecosystem has developed into a vibrant investment-intense system, with incubators working very closely with the entrepreneurs. I will also chronicle how the technopreneurs who received their profitable exit then took their lessons learned as serial entrepreneurs and have transformed themselves into Singapore's most active incubators and investors. Interspersed in the narratives, I have included for the academically minded interpretations of the opportunity-recognition framework of the technopreneurs, with use of the theory of effectuation.[7] Finally, in Part IV, I will describe two new ventures with very innovative offerings, having worked very closely with their mentors to raise the seed round funding and share changes they made to the business model and strategy. The two new ventures are presented in a teaching case format for the readers who are coaches, mentors, and educators to use.

This is also a book for technology entrepreneurs who are seeking funding and mentorship. It offers insights into the risks and opportunities of working (or not working) with your first investors. For anyone considering investments, this is a book about how to work effectively with stakeholders and manage their expectations.

Virginia Cha

Acknowledgements

In the middle of working on some long, knowledge-intensive project, I would often question myself, 'Why am I doing this? This is no fun'. But then, the project gets finished and I am filled with such a sense of gratitude (and relief).

It is finished; and I am touched again after rereading the stories told in the book. I am thankful to the many successful entrepreneurs who selflessly gave me their time and whose humility shone through in every single interaction and the humble words they used. I am even more grateful to the few who were willing to share their failed journeys – this is a very difficult endeavour in face-conscious Asia.

With this, I am mindful of the debts of gratitude towards the following (in no particular order):

- Jennifer Chi Chen Lien, who unwittingly joined the Institute of Systems Science as a research assistant, not realizing the challenge of having to manage me and my multiple, concurrent, over-committed, long-term, knowledge-intensive projects. We jokingly refer to our relationship as 'Freaky Friday', except that it is a freaky day, every day for her.
- Professors Wong Poh Kam of NUS and Desai Arasimalu of SMU who recruited me to be a mentor for incubated start-ups, which gave me numerous insights into the development of the technopreneur ecosystem.
- Professor Juzar Motiwalla of NUS who gave me the first challenge in Singapore – to commercialize technologies from the Institute of Systems Science.
- Lim Swee Cheang and Yum Hui Yuen, for being continually supportive of my comings and goings at the Institute of Systems Science (National University of Singapore), and who gave me generous support on the research portion on the construction of this book.
- Tan Ying Lan, Hugh Mason, Wong Meng Weng, and Ong Peng Tsin for their valuable contribution to the ecosystem in Singapore and also for the information. A special mention goes to Meng – whose creativity and insights set him apart among the many in the ecosystem.
- Dickson Gregory, Vincent Lau, Damian Chow, Xu Daxiang, and Quek Shu Yang for letting us write teaching cases based on their experiences in fundraising in the new ecosystem.

- Danny Chng, Danny Wilson, Chak Koon Soon, Ronnie Wee, Leslie Loh, and Ng Fook Sun – without your willingness to share so openly and genuinely about your journeys, this book would not be possible. Your stories are amazing and they need to be told. Jennifer and I are very proud to be the ones to write them.
- My mother, for being the role model she never intended.

Acronyms

A*STAR	Singaporean Agency for Science, Technology and Research
AA	Arthur Anderson
AC	Andersen Consulting
AIMS	Advanced Investment Management Solution
APAC	Asia-Pacific
API	application processing interface
ASEAN	Association of Southeast Asian Nations
B2B	business to business
B2C	business to consumer
BUNCH	Burroughs, Univac, NCR, Control Data Corporation, and Honeywell
CBD	Central Business District
CDN	Canadian Dollars
CD-ROM	Compact Disc Read-only memory
CEO	Chief Executive Officer
CFO	Chief Financial Officer
CNN	Cable News Network
COBOL	Common Business-Oriented Language
CTO	Chief Technology Officer
DBS	Development Bank of Singapore
DEC	Digital Equipment Corporation
DVD	Digital Versatile Disk
EDB	Economic Development Board
EDBI	Economic Development Board Investments
ERP	electronic road pricing
ESVF	Early Stage Venture Fund
FPGA	field programmable gate array
GE	General Electric
GLC	government-linked corporation
GUESSS	Global University Entrepreneurial Spirit Students' Survey

HBO	Home Box Office
HP	Hewlett-Packard Company
HTML	hyperlinked text markup language
IBM	International Business Machine
ICT	Information and Communication Technology
IDA	InfoComm Development Authority
IHL	institutes of higher learning
IIPL	Infocomm Investments Pte. Ltd
INSEAD	*Institut Européen d'Administration des Affaires*
IP	intellectual property
IPO	initial public offering
IPTV	internet protocol television
ISS	Institute of System Science
IT	Information Technology
JFDI	Joyful Frog Digital Incubator Asia
LAN	Local Area Network
LHIC	Lernout and Hauspie Investment Company
LHSP	Lernout and Hauspie Speech Products
MASS	Multilingual Application Support Services
MBA	Master of Business Administration
MDA	Media Development Authority
MIT	Massachusetts Institute of Technology
MNC	multinational company
MRT	Mass Rapid Transit
NASDAQ	National Association of Securities Dealers Automated Quotations (the US stock exchange)
NATAS	National Academic of Television Arts and Sciences
NCB	National Computer Board
NCR	National Cash Register
NEC	NUS Entrepreneurship Centre
NEI	NUS's entrepreneurship incubation program
NFIE	National Framework for Innovation and Enterprise
NOC	National University of Singapore's Overseas Colleges
NRF	National Research Foundation
NS	National Service
NSTB	National Science and Technology Board
NTT	Nippon Telegraph and Telephone Corporation
NTU	Nanyang Technological University
NUS	National University of Singapore
OCS	Officer Cadet School
OECD	Organisation for Economic Co-operation and Development

PARC	Palo Alto Research Centre
POS	point-of-sales
PSLE	primary school leaving examination
QR code	quick response code
RDBMS	relational database management system
RIEC	Research, Innovation & Enterprise Council
RM	Malaysian ringgit
ROI	return on investment
SAF	Sirius Angel Fund
SARS	severe acute respiratory syndrome
SDK	software development kit
SE	Southeast
SEC	US Securities and Exchange Commission
SME	small and medium-sized enterprise
SMU	Singapore Management University
SOC	School of Computing
SPRING	Standards, Productivity and Innovation Board
ST	Singapore Technologies
STS	Singapore Technical Services
T21	Technopreneurship21
TDF	Technology Development Fund
TIF	Technology Investment Fund
TIS	Technology Incubation Scheme
UK	United Kingdom
UNESCO	United Nations Educational, Scientific and Cultural Organization
USA	United States of America
VAX	Virtual Address Extension
VC	venture capitalist
VP	Vice President
WAN	wide area networks
WIMAX	worldwide interoperability for microwave access
YES	Young Entrepreneurs Scheme
ZTE	Zhongxing Telecommunication Equipment Corporation

Part I

Raising money for your technology dream

Importance of getting the first partner right

Technology venturing is hard. It takes a long time for technologies to mature, and it takes a lot of money. A typical process sees entrepreneurs pitching their ideas to win investments but too often they understate the difficulty of their business, especially when it comes to customer adoption. Their investors then become disillusioned with how long it takes the company to reach scalable growth. More attention needs to be paid to educate investors post-investment as this misunderstanding can be detrimental to a start-up's growth.

Virginia Cha (author), Danny Wilson, and Danny Chng are three technopreneurs who struggled to maintain clear lines of communication with their investors during the internet revolution of the mid-1990s. For every new technology wave, a new crop of innovators will emerge to disrupt the status quo in business, operations, consumption patterns, etc. As peas in a pod, every wave will also spawn the investors and financiers who are keen to capitalize on industry growth without truly understanding the technology field. How the technopreneurs choose to work with their investors makes all the difference.

1 Star+Globe Technologies and Virginia Cha

I came from a business family. Both my parents were entrepreneurs and owned a succession of mostly textile-related businesses. My mother, in particular, was always the one who had the energy and drive to start new businesses, and was not particularly constrained by lack of prior knowledge. She started a textile knitting factory in Bangkok in the early 1970s, followed by a medium-scale supermarket business in Hong Kong late 1970s, and then a cotton-processing and spinning factory in late 1980s in Bangkok. As typical Chinese entrepreneurs of traditional businesses, my parents' businesses were started with their own bootstrap capital and grown organically. Although I was exposed to business at a young age, the concept of external investors did not enter the family's vocabulary.

My earliest (failed) entrepreneurial experience was toting around a bag of leftover knitwear samples from my mother's factory to try to peddle to shops, businesses, and neighbors along Soi 1, Sukhumvit road where we lived. I was probably around ten years old at the time and was already practising effectuation (see boxed text for an explanation on effectuation). My parents did not ask me to do this. I took it on because I felt I could probably make extra cash from items that were lying around the factory. I tried to sell them to make a quick buck, but since knitwear was for cold climate places (my mother had an export business), needless to say I was not very successful. That was probably my very first experience in trying to sell a product that did not fit market needs. The funny thing was that I did not even need to do this to raise money. My parents had a fairly lackadaisical attitude for giving out cash allowance. I pretty much drew whatever I needed from a cashbox without any accounting. I suppose I did it just for the challenge. Or perhaps I was a bored ten-year-old.

Ours were typical pre-occupied Chinese business parents who paid scant attention to me and my two older brothers. We were pretty much expected to study hard and achieve a secure future through education. Interestingly, like most parents of this era, my parents did not see entrepreneurship as a meaningful path for us children, even if that was their chosen profession. The oldest brother was always expected to take over the family business, as was the tradition, but since I was the youngest and a girl, I was not expected to enter any particular profession or even have a career. All three of us completed our early education all the way through high school at the International School of Bangkok, a school attended by

Effectuation

The theory of effectuation offers a model to explain decisions logic taken by some entrepreneurs in the new venture creation process. The basic premise of this model is means-based (questions of 'who I am', 'what I know', and 'whom I know') thinking in turn derives possible effects from the means (Sarasvathy, 2008). Other theories to explain how entrepreneurs behave in environments where they make do with resources on hand include 'bricolage' (Baker & Nelson, 2005). Effectuation has a strong practitioner appeal because the effectuation framework provides an understanding of why and how entrepreneurs deviate from plans or often act without a clear plan. The highly cited theory has been enthusiastically embraced by scholars and educators in the entrepreneurship field. Textbooks on effectual entrepreneurship (Read et al., 2009) are available to educators to guide them on lessons on how large companies can learn from effectual entrepreneurship.

As can be seen in the extreme example of a 10-year old effectuating entrepreneur who took the opportunity to collect the left-over sample (means), armed with limited knowledge of trading (what I knew), and accosting the neighbors ('whom I knew'), means-based actions may not always lead to successful outcome, especially for novice and inexperienced entrepreneurs.

expatriate, primarily American, children whose parents were mostly in diplomatic or military service. I graduated from high school at the tender young age of 16, not because I was a gifted student, but because my parents skipped me through two grades (grade 2 and grade 5) to save money at the very expensive expatriate school. Consequently, I learned to think very fast and absorbed information very quickly to keep up with my classmates who were fully two years older. I was not a particularly social kid – actually I enjoyed being left alone and would spend hours engrossed in the Chinese *wuxia* (martial arts) novels,[1] in imaginary worlds where *youxia*[2] (loosely translated as 'wandering heroes') would fight evil for the heck of it, but never staying put in any one place.

In 1976, I set off for university studies in the USA, after winning the argument with my mother on why a Chinese girl from a business family should attend college. After all, in those days, daughters of business families were to be married off to sons of business associates and become *tai-tais*.[3] I always found it contradictory that my mother, who was the business go-getter in the family, would want a life of leisure for her daughter. My father was ever the more forgiving parent and was neutral on this issue. Nevertheless, I attended my first year at North Carolina State University at Raleigh and ultimately graduated from the University of Hawaii at Manoa with a computer science degree. Choosing computer science in 1976 was definitely entering the unknown, considering that there were no computers available in Thailand then. My high school aptitude test came back with excellent scores in logic and math, and the test recommended computer science/engineering as a field. I still remembered the geeky Jewish boy, Stuart, who sat next to me in the classroom: I asked him if he knew what a computer was also confessed that he too had never seen one. So, shrugging my shoulders, I ticked it as my chosen field. I guess that was how lifelong decisions were made back in the days before information proliferation, enabled ironically by computers. This turned out to be an excellent choice. Computer programmers were in great demand then. I remembered getting part-time gigs during college running COBOL[4] payroll programs for the College of Arts and Sciences at the University of Hawaii. Not only did I get to spy on my professors' salaries, I was paid a princely sum for doing so!

Always in a hurry to get on with life, I graduated, got married, and started my first job in the same month in 1980. In those early heydays of the computer revolution, you could secure great offers even if you were bored and nodded off during job interviews (which I did). I joined Burroughs[5] as an operating systems programmer for mainframes, and rejected (other higher paid and attractive) job offers from Fairchild Semiconductor, HP, NCR, and one other firm whose name escapes me. I chose the job at Burroughs because in those days, the best-known geeks were in systems programming (today, the geeks gain fame as hackers), not applications programming, and Burroughs then had the best architecture in systems design for mainframe computers. I had prided myself as a geek. I had a solid technical career at Burroughs, who later merged with Sperry Univac to become Unisys.[6] In 1984, I was promoted to run the operating systems group, and was the youngest, first female, and the first person of Asian descent in that position. By 1994, I had risen to the position of director of advanced technology for Unisys, and reported to the chief technology officer, in the prime of my fast climb up the corporate ladder. Around this time, this little-known phenomenon called the internet and html[7] was rearing its head in the commercial world. I became very intrigued with the possibilities of this technology and explored how we could port our mainframe-based development environments to one where the multiple R&D centers (and from two different mainframe cultures) could collaborate using open systems tools. I was able to convince top executives on the potential productivity gains from the adoption of the internet and we formed a small team under my advanced technology group to develop this idea. Meanwhile, within the company, debates were raging between the different development centers on whether to continue the mainframe business model or to adopt a service-centric model as the threat of open systems loomed and the bread-and-butter mainframe revenues were on a continuous and undeniable decline. In such a politically charged environment, it was nearly impossible to convince anybody to pay attention to new technologies emerging on the horizon, especially if they did not immediately intersect with our core business space. I found myself spending more time dealing with corporate politics and justifying decisions instead of doing what I enjoyed most – problem-solving with the use of technology. It was at this juncture that I decided to leave my first and only corporate job.

In 1995, the Institute of Systems Science, a research center focused on leading-edge information technologies housed on the National University of Singapore (NUS) campus but funded under Singapore's National Science and Technology Board (NSTB) recruited me to help commercialize technologies from its laboratory. Dr Juzar Mottiwalla was then the head of the institute and gave me the freedom to craft a commercialization strategy for the lab. Given a blank sheet of paper, I helped create the Commercialization Grant Scheme, which provided promising researchers S$500,000 seed funding to build a business plan while helping them search for external sources of venture capital financing. I thought effective commercialization of technology had to be done outside a research organization, and I was heavily influenced by the then-head of NSTB, the late Vijay Mehta, who did not think the government should be both the 'judge and jury' when it came to evaluating market opportunities for the technologies funded by it. Hence the entire scheme was to encourage spin-outs of the technologies with teams comprised of researchers and business-minded new recruits.

One of the ventures spun off under this scheme was co-founded by me while I was working on the various technology-transfer activities. The spin-off company built a set of applications and toolkit, named MASS, for encoding and decoding multilingual information based on technology research at the lab (see boxed text for detailed description on the technology). The young research team (all non-PhD research engineers) from the lab – Ho Yean Fee, Wilson Lee, Li Jian Zhong, Chong Chiah Jen – after some coaxing agreed to leave the safety and relatively stress-free environment of the lab to join the first venture-funded spin-out.[8] The PhD researcher and software architect Dr Ngair Teow Hin declined to spin out with us because he had wanted a different commercialization path for MASS. He had eschewed venture funding and wanted to grow the business via a combination of bootstrap capital and customer-driven revenue via solution integration consulting services to large corporate customers. At that time, in the mid-1990s, venture capital funding for seed-stage software development companies were virtually non-existent, so this organic growth path was the most common choice for funding a technology business in Singapore. In hindsight, as events would evolve, Teow Hin probably had the answer right for MASS. But, I was convinced that we needed large capital investments to develop MASS properly to win in the global market. Perhaps because I had spent my entire career up to then in a Fortune 500 company and was used to reviewing multi-million dollar development budgets, I did not think the bootstrap capital path was worth considering if the technology needed to be developed to be the leader in the fiercely competitive global market.

Concurrent to my commercialization effort, NSTB had formed Singapore's first seed-stage technology venture fund, aptly named the Technology Development Fund (TDF) and modeled the investment fund on Silicon Valley's venture capital fund management. In parallel, Singapore's first commercial

MASS (Multilingual Application Support Services)

MASS was the internal project name for this development effort while in the research lab. Amazingly, the team and the development efforts for MASS originally did not envision it to be a 'big idea', ground-breaking research agenda. MASS evolved since 1989 because the research staff needed to create a set of supporting toolkits to ease their own development efforts in the multi-language processing research program. The toolkit basically enabled processing of the Unicode (universal code) character set, and contained input and display modules and a collection of information-processing tools in the form of APIs (application processing interfaces) to enable searches, queries, and manipulation of Unicode strings on both the Unix and PC Windows platforms. Competing technologies offered bilingual products – this meant they could support one additional script other than the Roman script, which meant that developers/users had to purchase multiple toolkits if they needed to support multiple, non-Roman script languages. Because MASS was designed to be a multilingual system, the architectural design was necessarily extendible and customizable. Hence, addition of new language support was relatively simple in MASS. This enabled a small team of software engineers and a PhD software architect to develop and evolve the toolkit based on a combination of internal requirements and perceived external market potential. Singapore Airlines became the first commercial customer with the sale of one licence in 1992 for the development of the multilingual menu-planning application system. In 1993–4, two more licences were sold. By 1995, MASS caught the attention of many application developers in various sectors and this resulted in over 30 licences worldwide, while the technology was still in the research lab. The major application was the library sector, with National Library of Australia, McDonnell Information Systems as reference accounts. A typical software development kit (SDK) licence for the Unix platform was priced at S$20,000. At that time, the operating systems in the market (Unix, Windows) did not support Unicode as it was an emerging standard. MASS was the world's first multilingual information-processing toolkit and was fully two years ahead of all competition, including the platform vendors.

internet provider, Pacific Internet under Sembawang Media, was formed to look at new opportunities from the internet space. I pitched the multilingual technology business plan to TDF and Pacific Internet and secured the first S$2 million cash funding in 1997. Together with the technology holding arm of ISS, which held shares in exchange for the technology licence, Star + Globe Technologies was formed with an initial paid-in capital of S$3 million ($1 million was in-kind contribution for the technology IP) with three equal shareholders. The cash investors agreed to fund the company only if I accepted the job as CEO and leave the institute to run the company full-time. There was a carve-out of a stock option pool of 15 per cent for the co-founders and employees. In those days, we were not very sophisticated as co-founders and did not fight for a larger pool for founders. As a consequence, the founding team left the institute and, collectively, we actually did not own any shares in company. This skewed shareholding, where the investors held the majority shares while the founding team only had minority options, would become relevant later in the company's development. Nonetheless, I did not hesitate and left the institute, even with a pay cut, to be the first CEO of Star+Globe Technologies, thus ending my relatively short career as Singapore's de facto first technology transfer officer. It seemed like the right thing to do at the time. After all, I had convinced everyone about the viability of the company – from the investors to the founding team who also left the institute.

Our technology was world-renowned. We were approached by developers from all over the world to inquire about the technology. We had garnered many technology awards, including the prestigious National Infocomm Innovative Product Award in 1998; the co-winner for that award was coincidently Leslie Loh from System Access, now the managing partner for Red Dot Ventures, the new technology incubator supported by the National Research Foundation (NRF) of Singapore. The relevance of Leslie Loh and Red Dot will appear several chapters later in book when we talk about successful technology ventures and the new generation of incubators run by technopreneurs. Even with the accolades, we still had to fight for every customer to gain momentum and growth because Singapore companies were not keen to adopt leading-edge technology from a small start-up. Within two years, we were moderately successful in growing revenue from zero to S$3 million as a multilingual technology tools supplier to primarily international customers like Sybase, Fujitsu, GEMplus, etc. We even made inroads into banks like Citibank, JP Morgan, and Banc Paribas. Our only Singaporean corporate customers were Singapore Airlines and the National Library Board. We had successfully raised a second venture round at double the initial valuation, and attracted venture capital from leading firms like Vertex Management, Walden International, NIF Fund (a Japanese fund), in addition to top-up of investment funds from TDF. Running a technology product company was a difficult business and even though we had healthy revenue and customer references, there was always this sense of insecurity about our future. Star+Globe was chasing the technology frontier in offering multilingual information-processing tools by plugging the gap on popular operating systems platform such as Windows and Unix. As soon the market need became obvious, the operating system providers would develop similar toolkits and embed them into the system. Hence, we always had to develop the next generation of tools, which meant that we were always developing for the early adopter market, a niche and hard-to-reach market segment. We knew we needed to transition from offering only tools for early adopters to more mass market oriented applications and had plans to develop such applications like multilingual search and database query tools. But this transition took time as we had to start with the technology building blocks that we created. During this transition, we earned cash flow with a healthy solutions and consulting business, which also gave us insights into future market needs. The difficult dilemma here was the balance in resource needs between the product development team and the solutions integration team. Every resource taken from the product team to help the solutions team would delay the development of the standard product; on the other hand, starving the solutions team would deny the company the cash flow. This was actually the most difficult dilemma for Star+Globe. Because I came from a product development background and because a product-centric business has quantum leaps in scalability, I leaned towards resourcing the product team. Every custom project was evaluated against the larger goal of creating a standard product and we would turn down projects that deviated from this goal.

In 1999, TDF had introduced a partner European investor with whom they had co-invested in other Singapore-based start-ups in the speech and language-processing space. The European investor, Lernout and Hauspie Investment Company (LHIC), was interested in taking a stake. At that time, our company was on track to be cashflow positive. We had revenue, large corporate customers, and a dedicated team of 20+ people. We also established a solutions office in the USA and a sales office in London. We never used the cash from the second venture round – so we had over S$4 million in cash, a princely horde for a young Singaporean start-up less than three years old. By then, we also had a fairly deep product line in multilingual technologies – from multilingual information storage, retrieval, and display to search engine, complete with query-processing capabilities. Our technologies were not consumer-based applications – we offered these technologies in the form of an SDK or through APIs, so our customers used them to build their own applications. The sales cycle of our technologies typically took three to six months and involved buy-ins from the financial buyer, technical buyer, and the end-user from the customer's organization. Consequently, we were at odds with (some of) our shareholders' impatient expectations of being a hot new start-up in the ever-expanding internet dotcom bubble forming in 1999. I remember having to spend many meetings to educate the lead investor. He had wanted us to transition to be a dotcom, the in-vogue term at the time. Every company was trying to be a dotcom and he had pressed for an e-business strategy from our developer-centric Star+Globe. I had proposed a strategy to create an 'MLWizard' on the web where it would be the developers' community watering hole for information, tools, hints for multilingual development built on a freemium subscription business model. This would enable us to transition from selling individual licences (and their long sales cycle) to focus on transferring knowledge and mainstreaming multilingualism, but still remaining true to our original vision and target market of software developers. The lead investor did not understand why we could not target the large consumer market pool to achieve faster revenue growth and customer adoption. In those days, the idea of building a tech-community was not an exciting story. He never understood our technology – but he had a mental model of success for technology-based ventures, and that pattern read like a Silicon Valley story. In fact, in one of my many frustrating conversations with him, he rambled on about how we needed to recruit a US CEO so we could get traction in the very important US market. I recalled being mildly offended since I am an Asian American with a blue passport. The conversation went something like this:

INVESTOR: ....and we need an American CEO...
VIRGINIA: But I'm American.
INVESTOR: Oh, no, not you, Virginia. I mean, a real American.

That was how some Singapore-based investors looked at technology venture talents in those days. We had revenue and were profitable, and we had a lot of money in the bank so we were ideally situated for a backdoor listing as envisioned by our very impatient lead investor.

Of our investor group, TDF had a number of co-investments with LHIC on other companies and wanted to see how they could accelerate Star+Globe's growth through a strategic partnership. LHIC had previously invested and owned shares in a struggling company named Accent Software, which was trading on the US (secondary stock exchange) bulletin board, just been renamed Languageware, and was in its final year before facing the possibility of delisting. TDF and LHIC hatched a reverse-merger plan where Star+Globe would be folded into the listed entity, which was later renamed to WholeTree.com. The other company needed our cash and our technology base, and our shareholders were convinced by the lure of a quick win and a place on the dotcom wave, so this plan was perceived to be a win-win for the investors of both companies. Languageware had recruited a US executive from Sony, a true-blue American, to be the CEO of the company. The Languageware deal also followed the Silicon Valley script book and a retired grey-haired advisor was recruited as chairman for the company. I was not enamored of the reverse-merger plan because I felt the Languageware CEO had a completely wrong go-to-market strategy for the multilingual technology. He wanted to turn the merged entity into a B2C company, offering language translation services with use of our tools as a productivity environment. Recall that our technologies were built for multilingual information-processing, and did not actually perform language translation – an entirely different animal when it came to language-processing technology. In addition, our technology was not suited to support a B2C business model; that would not be a simple pivot in strategy – it would be a strategy leap. Nevertheless, I was powerless to stop the merger event as I was a minority shareholder holding only 3 per cent after my stock options had vested. The reverse merger completed in December 1999, four months away from the height of the dotcom bubble.

By now, the merged entity had many people at the top (the CEO, the chairman, and many well-paid vice presidents of assorted marketing and sales titles) who did not understand what business we were in, or the technology foundation for that business. They perceived the lack of fast growth as a marketing issue so proceeded to invest millions in an elaborate marketing campaign, which included advertising on primetime national television in the USA. The company even had commissioned a jingle, sung by a very cute, boyish-looking actor, toting the benefits of multilingual technologies. It was madness. It was painful for me to watch as I saw it as a cash-burning exercise of the hard-earned cash from Star + Globe. But for a short while, this barrage of publicity did work magic at the height of the dotcom froth period and the stock price shot up and gave the merged company a valuation close to US$95 million from near zero (recall the company prior to the merger was a living-dead shell trading on the bulletin board) within a span of three months. The meteoric rise in stock price enabled the company to raise additional working capital from secondary offerings, and more millions from non-technology investors were pumped in. In those heady internet bubble days, everybody was dumping their life savings into buying dotcom stocks for the quick returns. In Colorado where Languageware was headquartered and which was home to a number of US airlines, even pilots poured their entire cash holdings into the dotcom craze. At any case, I was long gone by then. I had been fired in early January 2000, less than two weeks after the merger completion and right before the very first post-merger management meeting. In Chinese, we have a saying, 'Two tigers cannot occupy the same mountain' (山难容二虎) and as I had disagreed on the direction of the merged company, naturally I was asked to leave. Since I was no longer with the company, I did not have visibility to what transpired in the company after my departure, other than what was published in the SEC filings. What I did manage to dig out, post-event, was that Languageware was seriously in debt prior to our merger, and LHIC was a majority shareholder. The merger was an event that triggered a prior contract signed in mid-1999 that enabled more money to be invested into the company from other investors. Without the merger, Languageware was in serious doubt as an on-going concern. All these unsavory facts emerged in various news articles and court documents many years after the merger. Sadly, all this is irrelevant to what happened to Star + Globe, a once promising technology star performer from Singapore.

In May 2000, the dotcom bubble burst, which had a sobering effect on companies that did not have a cashflow-positive business. Wholetree.com had burned a lot of cash on marketing, which had a short-term positive effect on the stock price but without the requisite customer revenue results because fundamentally, the business of language enablement of information processing is a technology commitment and each corporate customer had to be won through an elaborate technology evaluation cycle that took three to six months. The MASS technology helped enablement but did not perform actual machine language translation no matter how much the executives wished for that magic solution. Shortly afterwards, Wholetree.com was delisted and the staff disbanded. It became a non-operational entity and the IP was sold off. The technology lead was lost and 12 years of R&D investments went down the drain. To many in Singapore who did not know the inside story, Star + Globe may seem to have been

Lernout and Hauspie Speech Products (LHSP)

LHSP was a Belgium-based technology company with offerings in speech-recognition, language-translation and processing tools. The company was founded in 1987 by Jo Lernout and Pol Hauspie and later, in 1995, listed on NASDAQ. The two founders (among others from the company) were later convicted of accounting fraud and reportedly jailed. Lernout and Hauspie had set up an investment vehicle, Lernout and Hauspie Investment Company (LHIC) in 1998 to acquire small technology start-ups in the language space.

One of the dubious accounting practices that ultimately led to the collapse of LHSP was the use of related company transactions that were booked as revenue for LHSP, but were not properly disclosed. The most flagrant transgressions were for inappropriate revenue reported from Asia (Korea and Singapore were the two sources of inappropriate revenue reported in the business news). The fraudulent reporting of inflated revenues lifted LHSP's stock prices during the dotcom boom, which then allowed LHSP to acquire more companies, using their NASDAQ-traded shares as currency. At its peak price in March 2000, LHSP had a market valuation of approximately US$9.3 billion.

One of the more spectacular acquisitions, Dragon Systems Inc., a US-based company founded by two leading researchers, was acquired for reportedly US$580 million in LHSP shares in 2000. At that time, Dragon Systems was the technology leader in speech-recognition technology. The acquisition of Dragon and Dictaphone Corporation, another US company, necessitated fuller SEC requirements disclosure of revenue reporting by geographic region. The substantially high revenues from Asia trigged an investigative news report (and, in parallel, an SEC investigation), which was published in 2000. In August 2000, LHSP was forced to restate her revenue. The company was declared bankrupt in 2001.

a success story, having achieved public exit on NASDAQ within three years of founding (via a backdoor listing). Certainly I did not suffer from that particular exit as I became a hot tech talent shortly after my separation from Star+Globe and was inundated with many lucrative offers immediately after my separation. It was an exhilarating journey, but Star+Globe's story was more instructive as a cautionary tale than a success news blip. There is no short-cut to building a technology business. To build a successful technology business, you have to win the early customers one by one, and refine your products to meet the evolving needs of your targeted segments. Your business scales because your technology meets the requirements of your (hopefully) growing market segments and your technology is one of the few market suppliers to fulfil this demand with a superior offering at a profitable rate to your company. No amount of financial maneuverings can short-circuit this process.

It is unproductive to dwell on counter-factual thinking on what might have been. Rather, reflecting on my first start-up experience deeply imprinted on me the importance of getting the right first investor for a young technology venture. Often, entrepreneurs take money from investors because they are willing to invest at the agreed-upon valuation. The entrepreneurs need to see the investment as the beginning of a long-term relationship where attention and effort must be invested from both parties. In my case, I did not succeed in educating the investors on the intricacies of our technologies, nor did I think it was important to input the time or interest in bridging this gap in understanding. During the investment courtship stage, I did see hints of misunderstanding, but I did not bother to take the time to educate since my goal was to raise money quickly so I could get on to the real business of running a technology company. This implicit complacency in allowing the investors mentally to develop a more optimistic model for the venture is likely committed by almost all entrepreneurs in our quest to close investments. The only cautionary note for entrepreneurs is to ensure that your investors develop a more realistic picture on the nature of your business. Failure to do so can result in interventions by investors, which can completely torpedo the business in the worst case or, in the best case, completely sideline your investors where they cannot help you progress, leaving you to fight a very lonely war.

2 Pixelmetrix and Danny Wilson

Tourists visit western Canada to absorb the beauty of the natural landscape. From the rugged peaks of the Rocky Mountains to the emerald green waters of Lake Louise, there are an endless number of sites perfect for ski vacations and camping weekends. The clean air and socially progressive atmosphere draws in countless young professionals and immigrants seeking to claim a piece of this pie every year. In 2011, *The Economist* named Vancouver the most livable city in the world. Prized as a city with the best of both worlds when it comes to beaches and mountains, it is also a favorite of singer Bryan Adams and LA Lakers point guard Steve Nash, who often spend time in the beautiful city to unwind.

One thousand kilometers northeast of Vancouver is the oil-rich city of Edmonton. The Betty to Vancouver's Veronica,[1] many are drawn to the city for the practical reasons of growing employment opportunities, especially in the oil and gas industries, and low taxes. Housing just over one million people, Alberta's capital city has a cozy, small-town feel. While Vancouverites and Edmontonians are both proud of their identity, the former is flashy while the latter prefers understated chic. Because of this, Edmonton's tagline 'City of Champions' might appear ill suited. The brazen statement was assigned by then mayor Laurence Decore in the aftermath of an force four tornado that saw citizens come together and rebuild the city. Neighborly and reliable, Edmonton continues to be a lovely place to raise a family.

Growing up, Danny Wilson enjoyed all Edmonton had to offer. Like many other middle class Canadians, his mother was a homemaker and his father worked as an airport radio navigation engineer with the government. A hands-on, technical, 'get it done' kind of man, his father always had numerous projects on the go at home, and this environment rubbed off on Danny as he too started on numerous projects in electronics and chemistry from elementary school. Of course, an understanding mother was also important – especially when one experiment filled the entire house with white smoke so thick you could not see your hand in front of your face. Aside from hobbies in science, Danny actively participated in the school jazz band as well as the Edmonton Youth Orchestra. While considering (very) briefly a career in music, the die was inevitably cast from a young age: Danny would pursue some kind of technical career.

But which career? His mother thought it would be a good idea if he took up a stable profession, as his father had. While Danny was not entirely sure what he did want to do, he was sure he did not want to work for the government. He coasted through high school, getting high marks with minimal effort, and upon graduation in 1981 decided to stay in his hometown to attend the University of Alberta to take up a degree in computer engineering.

There is a running joke that if you bump into a Canadian, they will apologize to you for being in your way. This humble and congenial nature typifies Danny. When meeting him for the first time, most people are struck by his unpretentiousness and honesty. He tells it like it is and stays true to himself. In school, he avoided self-serving student leader positions preferring to play in the jazz band and teach community classes on the fundamentals of computers. While teaching did not become a passion, he loved learning about the ins and outs of the engineering. The next summer, he found a position at a small engineering firm and got his feet wet doing some advanced programming. As the firm was small, and Danny was both motivated and capable, he was assigned a wide variety of projects in both hardware and software design. While engineers in large companies are usually confined to a narrow discipline, the broad exposure to different areas of engineering gave Danny a unique perspective.

The firm was like a close-knit family working hard, long hours on cutting-edge technology and playing hard whether at a company-sponsored dinner, barbeque, or slow-pitch softball against the other electronics companies in the area – where both teams always ended up at the local pizza joint! He asked for, and was granted, a full-time job as an engineer after graduation and he settled into a comfortable routine as a young professional in Edmonton. The engineering firm he worked for specialized in niche test and measurement equipment, and his first job was to write a specialized field programmable gate array (FPGA) compiler. Life quickly became exciting as he loved engineering high-tech products. His company secured a deal to provide advanced test equipment to NTT of Japan, the largest telco in the world at the time. In those days of Japanese protectionism, it would be the first purchase of that kind of equipment from a foreign company. An important and sensitive deal, Danny was sent to Japan to test the system because of his thorough understanding of the nuances of both hardware and software.

Shortly afterwards, he was promoted to the customer support division for North America and Asia. For two years, he worked as the manager and editor for the technical manuals and documentation group and later was promoted to product manager for new product development. It was in this role that he found his 'sweet spot'. What he loved most of all was talking to customers, translating their requirements to products, and using an imaginative understanding to cater to their needs. He was soon given the opportunity to spread this passion to a wider audience when his firm was acquired by HP. Danny thrives in small entrepreneurial atmospheres and the acquisition could have posed a problem for him, but the firm did not turn corporate. At least not right away. For a while, they maintained a small-company feel receiving few visitors from corporate honchos due to Edmonton's remote and bitterly cold location.

Big firms do have their advantages, and despite rising to division marketing manager and having lived in Edmonton his entire life, he decided on a change of scenery and accepted an opportunity to start a new group in Kobe in Japan, leading a small R&D team tasked to build applications on one of the products he had made previously. A short trip from the Osaka mega-city, in some ways Kobe rivals Edmonton especially when it comes to its medium size and strong reputation as a fledging research city. Where Kobe comes out ahead is in the areas of history and culture. Designated as a UNESCO City of Design in 2008, the city is known to many as the 'Paris of the East' and is a cosmopolitan port city. Original excitement about the new city and new challenge quickly turned to concern as the Great Kansai earthquake hit the city as he was preparing to move.

Despite the challenges of navigating the rubble and multi-year road constructions for five years, this environment was an inspiring place for Danny creatively, and contributed to his success at work. Minus a few bureaucratic frustrations, he was happy working for HP and loved his adopted city. This all changed in 1998 when HP decided to spin out Agilent Technologies. For one year following the announcement, there was what Danny calls a 'huge divorce' where the assets were being divided and Danny's division was caught in the crossfire.

The entrepreneurial-feel in HP was gone, customer-focus was fading fast, and Danny found this new atmosphere stifling. He decided to pivot and apply for an MBA program, thinking the new environment would naturally open doors that would take him in a new direction. With his arsenal of intricate technical and cross-cultural experience, he was accepted at MIT's Sloan School of Management. During this time, a customer at work approached him with a challenge. His client, DirecTV, needed an instrument that could test 200 satellite channels. The model offered by HP could only test one, and it was not economical for him to purchase 200 of them. 'Could you make that?' he asked.

While Danny is naturally low key, his competitive streak comes out when challenged. As an ambitious engineer, his gut response was 'Of course I can do that'. His gut told him two things: that he could direct a team to complete the project and that if this customer wanted the device then it was likely others would as well. Approaching the division and group managers, he explained: 'Look, we have this requirement from one of our customers. This will require some development effort but if we make it, he will buy US$2 million dollars' worth of equipment'.

For months, they remained hesitant naming the preparations for the Agilent spinoff as a higher priority. All the while the customer asked every week 'Can you make it or not?' Sick of the chaos, Danny prepared for the move to Boston.

Things changed when an old friend from his university days came to visit. Ben Lim was a former classmate of Danny's at the University of Alberta and had since then been working for a number of high profile banks, most recently running the global computer systems for one of them. While out drinking one night, they bonded over frustrations pent up over years of working for big firms and dreamed about becoming entrepreneurs. Their conversation was the turning point for Danny, who soon after decided he could do this. He briefly pondered starting the business while studying at MIT; but this suggestion was quickly squashed by Ben who agreed to quit his job if they were to give entrepreneurship a real shot. Danny agreed to focus fully on building a business and, at Ben's suggestion, decided to move to Singapore.

Singapore, colloquially known as the Lion City, is one of the world's best places to do business.[2] Geographically near some of the world's most sophisticated technology and manufacturing markets, it is home to a stable government, educated workforce, low corporate income tax, and English fluency. In the early 2000s, it was also more affordable than their other preferred options of Vancouver (high taxes), Tokyo (lack of English-speaking engineers), and Silicon Valley (exorbitant engineer salaries). They had also heard through the grapevine that in Singapore it was easy to hire quality engineers cheaply, though this turned out to be untrue. The cheap engineers were just not that good, and the quality guys were not around. Primarily this was because the start-up scene at the time was very primitive and working for one did not have the cachet of working for a multinational company (MNC). What Singaporeans valued most of all was obtaining the five Cs of cash, car, credit card, condominium, and country club. Working for a big firm meant job security and the potential to move up. Danny and his team had a lot to overcome, and he describes the difficulties as follows,

> In Pixelmetrix we give you a blank sheet of paper and ask you 'What are you going to invent today?' We need to create exiting technological products for our customers in Los Angeles, New York and Sydney. And it's funny that if I offer that value proposition to an engineer in Vancouver and Silicon Valley they would jump on it. In Singapore they would give reasons like 'My mom never heard of this company and it doesn't have a brand name. I would rather work for IBM as a support engineer.' It's a huge barrier to being successful in this place.

Danny and Ben recruited Tom Orlowski, a hardware architect from Vancouver, and Hideki Takahashi, an engineer from HP Japan, to complete the start-up team. Together they formed Pixelmetrix, a firm focused on high technology products for the television sector.

If it sounds strange that a telco professional made the jump into television, rest assured Danny too is surprised by the journey, as well admitting 'Technically, digital TV is not so different from telecom – both send content over packetized networks so it should be easy'. However, he quickly realized that was not the case.

The telecom industry is dominated by a small number of massive players, with each country hosting one or two big companies. In China, there are Huawei and ZTE. In Japan, there are Fujitsu and NEC. In Europe, there are Alcatel, Siemens, and Nokia. The television industry, on the other hand, is comprised of thousands of small companies – each providing a small part of the overall system. Unlike the telecom giants who provide a wide range of equipment, television equipment providers are highly specialized and often provide only one type. Furthermore, for each, there were usually only a handful of suppliers globally. 'Think about all the

About Pixelmetrix

Pixelmetrix specializes in preventive monitoring for digital television networks. Our array of Test, Measurement and Monitoring systems define a new generation of tools and techniques for the quality control of broadcast signals and content.

Pixelmetrix provides network intelligence implements and systems to television broadcasters and telecom operators for the management and monitoring of broadcast quality of service and quality of experience. We tie technological expertise from the broadcast, telecom and IT sectors to provide innovative, award-winning and cost-effective solutions worldwide.

Our products employ parallel processing, highly scalable architecture which enables service providers to move beyond today's manual, human-based monitoring of service quality to a fully automated and distributed large-scale monitoring system.

Pixelmetrix is a member of DVB, NAB, IABM, SCTE, SMPTE and the ITU's Video Quality Experts Group (VQEG). The company was founded in May 1999 and is headquartered in Singapore.

Source: Pixelmetrix website. Reproduced with permission.

"stuff" you need for television', explains Danny. 'You need microphones, tripods, cameras, lenses – this list goes on and on and on. In each category, there are only four or five companies. It is quite a closed and intimate "club"'. Despite the new environment, he was determined, as CEO, to carve a niche for Pixelmetrix in the field of preventive monitoring.

That said, the journey towards success was rocky. The team had an uphill journey, needing to navigate unforeseen roadblocks and to educate their engineers and investors alike on their niche market. Their first task was to fulfil DirecTV's US$2 million contract. When Japanese companies purchase anything, approvals are given using the '*ringi*' system where a document is circulated around to all parties involved in making the decision and each layer of management gives approval by stamping it with their unique stamp. Amendments and comments are added throughout. Depending on the value of the purchase and amount of bureaucracy, two to five stamps are typically required. For the DirecTV contract, five stamps were needed, one each from the engineer, engineer's manager, section manager, group manager, and CFO. Only once all five stamps were in place would the company release the purchase order. Unfortunately for Pixelmetrix, with only two more stamps to secure, Takahasi approached Danny with the news that DirecTV had been acquired by their closest competitor who had decided to close the facility and fire all the staff. The financial security they had hoped for with the contract vanished and the team had to search aggressively for additional customers. For this, they had confidence that they would be able to find customers once the product was released.

Called the DVStation, their first product was a device capable of quickly identifying and isolating errors in video transmission chains, which would appeal to the many television providers switching from traditional to digital broadcasting. Manually monitoring systems and uncovering errors was time-consuming, and as operators everywhere were migrating their analog systems to digital, their product filled a gap in the market for tools that could take over these responsibilities. However, luck was not on their side as, in an ironic twist of events, the new product was launched on the same day as the NASDAQ crash of 2000, which severely hit the IT and broadcasting sectors. This left Pixelmetrix with a developed product that few customers had the funds to purchase. Danny and his team had been determined to self-fund Pixelmetrix for as long as possible, buying all the office furniture secondhand, cleaning the office themselves, and even cashing in their retirement savings, so their valuation would be as high as possible. They also accepted some angel money. But the purchasing delays from the crisis and the resulting financial stress on Pixelmetrix forced them to court new investors.

In early 2000, there were a number of networking events in Singapore for entrepreneurs to meet investors and pitch their ideas. A straight shooter, Danny avoided smarmy small talk and tried to win over investors by impressing them with the sophistication and technical elegance of Pixelmetrix's products. Looking back it was unsurprising that the first 100 or so pitches that he gave were unsuccessful, as most investors did not come from a technical background and glazed over when Danny spoke even briefly about the technology. In those twilight days of the dotcom era, 'I was always asked "what's your business model?"', said Danny, 'and they were always so surprised when told that Pixelmetrix made hardware and software, marked it up, and sold at a profit. Few investors expected or understood a "brick and mortar" business'. Changing his pitch to focus on the products' potential disruptive impact on the television industry's move to digital, he was able to attract many interested investors. Many of these were supportive and promised to link Pixelmetrix with their 'extensive network of contacts'. Soon they had over a dozen investors and plenty of cash, which resulted in a different problem all together.

The start-up industry was getting hot and many venture capitalists (VCs) were keen to get in on the game. Comments in board meetings would go along the lines of 'So can you invent some product so the company can get S$10 million revenue by September?' Influenced by the success of well-known mobile games, search engines, and music players, some of their VCs wanted Pixelmetrix to piggyback on their success and earn quick cash. For the preventive-monitoring industry, it is impossible to make quick revenue. Much of the work is custom orders and customers will not purchase unless they can see a prototype. The development from conceptualization to the final purchase runs on a three-year cycle: first Pixelmetrix produces a prototype; then they show the prototype to a potential customer and customize it for their needs; and finally the customer will put it into their budgets for the following year. The early board meetings were a frustrating process for both sides. Danny explains his early frustrations as follows:

> One of the things about VCs and I observe this about them from the board meetings – listen to what they are talking about, they don't talk about the same things you do. They live vicariously through the success of others saying, 'This friend of mine invested in that XYZ Company which makes that little thing that attaches on to the whatever and he made 7 billion dollars,' or 'Oh yeah, well I know this guy that made an investment into ABC Company. They went IPO and now they have 11 billion dollars.' And they all repeat it again – which gets very tiring. They say again and again and again how much money their friends are making. 'You know, one of my friend's partner's companies made 50 billion dollars making a search engine. You have software engineers right? Can you make a search engine?'

While this description is obviously a tad exaggerated, it showcases the problems that can occur when entrepreneurs and investors have different visions of what path the company should take. Over time, Danny learned the secret to success when dealing with their board of directors: communication. In their case, it was initially difficult for their investors to bring in real value and share their network effectively because they were not on board with Pixelmetrix's vision. In the start-up world, angel investors and VCs often invest in areas where they have little expertise. Sometimes, this is because they have previously done a successful deal with another investor on board; other times it is because they follow a lead investor who foresees growth in the field. With some effort, entrepreneurs can educate their investors – whatever their reason for investing – on their vision, heed effective feedback, and work to progress the company. To do this, entrepreneurs need to have a clear vision for the business.

Danny had a clear vision for Pixelmetrix, stuck to his guns, and eventually the success of his products spoke for themselves. Today, the company is a leader in the terrestrial, satellite, cable, and IPTV (internet protocol television) network industry and their client list includes television giants such as Viacom, Disney, CNN, and HBO. Adding a bit of glamour to their industry, they were the recipients of Singapore's first Emmy award from the National Academic of Television Arts and Sciences (NATAS) in the category of Engineering and Technology for their DVStation Transport Stream Analysis products. They have also gained a strong following for their Electronic Couch Potato for remote viewer quality monitoring and consolidator series of network management software. Budding start-ups should take heart that a difficult relationship with investors does not necessarily prevent a transformation into a successful company.

While Danny's investors gradually came to accept his vision, they never truly understood the nature of the preventive-monitoring business. Still today, many focus on chasing the next big thing with some continuing to suggest Danny chase their leads. The biopharma segment and China's services sector are top picks with one saying, 'Hey, maybe Pixelmetrix should do more services in China'. By refusing to give up on his vision, Danny was able to overcome early missteps to find his footing. Today, although they are not involved in what Danny jokes as being the 'glamorous movie star' side of television, they continue to play an essential role in powering the industry.

3 Sinetics Associates and Danny Chng

In the highly competitive symphony that composes southeast Asia, education is viewed as a good predictor of future success. Students in developing nations are especially keen on excelling academically, seeing schools as the channel through which they can obtain a better life. Over the years, this environment has attracted many private education providers, from elementary to tertiary levels, keen to capitalize on this interest with varying degrees of repute. Malaysia is one country that is working to shut down disreputable organizations while inviting prestigious global institutions to set up campuses in EduCity, their new 600-acre educational haven of global universities, R&D centers, international schools, and conference centers. Marlborough College, the famed British independent school also known as the *alma mater* of Catherine, Duchess of Cambridge, chose Malaysia for its first international outpost. Their vision of preparing globally competent students matched EduCity's vision, and they set out to build their campus from the ground up. Set to open in the Autumn of 2012, the Marlborough Malaysia campus has a similar feel to the Wiltshire original with stately buildings and 90 acres of rolling green fields. The school will bring a bit of international glitz to Malaysia's education sector, but this does not mean that students who do not attend storied private institutions are relegated to a second-class education. There has long been a selection of elite public institutions that have trained the region's highest-ranking public servants and business elites. Alumni of schools such as the Victoria Institution includes Ananda Krishnan, Malaysia's second richest man, and Chung Ling High School, from which hails several of Singapore's past ministers, top government civil servants, as well as Khaw Boon Wan, Singapore's current Minister of National Development.

Danny Chng was one of these Malaysia-born and bred education superstars. As a varsity athlete with straight As at the top high school in the country, he had the kind of confidence typical of an accomplished student leader. In a society where peer pressure referred to academic pressure to score at the top of the class, he had that in the bag. What set him apart from his peers were his rebellious streak and his tendency to challenge the status quo. Some might say he was a responsible rebel, not going out of his way to break the rules but always finding ways to circumvent the ones he did not agree with. One way he won over his peers was by successfully pulling off a series of what he calls 'harmless' pranks by outsmarting others. He recalls:

> We were not truly destructive. We consciously knew what we were doing. For example, when we painted the windscreen of our teacher's cars, we made sure we used water-soluble coloring, similar to what's used in making edible Chinese cakes, so it could be washed off. Teachers never knew who did this and we were never caught. Our classmates that knew of these pranks looked at us like we were chiefs and admired our guts.

This natural leadership culminated in his election, by both fellow students and his teachers, to the leadership position of prefect. Students are typically chosen to be prefects based on their leadership qualities, maturity, and ability to evoke respect among peers. Or, according to Danny, 'If you have a good academic record and if you are naughty, then you get to become a prefect. We knew it was the disciplinary and headmaster's strategy to keep us out of mischief'.

During his school days, Danny's ability to take big concepts and explain them to the masses using simplified terms showcased his ability to think deeply. He was brazen but contemplative; his parents trusted in his ability to make his own decisions. The eldest son from a typical middle class Penang family, he was expected to be a good role model for his younger twin brothers but was otherwise free to choose his own path in life. His father, a bank worker, and his mother, a schoolteacher, were generally satisfied if he brought home strong grades. He did, and upon graduating from high school in 1978, he moved half way around the world to make his mark on a new environment. In contrast to Malaysia's tropical milieu, he chose to settle in the provincial Canadian city of Edmonton to attend the University of Alberta, at the time one of Canada's top five universities, and obtained a degree in computing science during a time when most people did not know nor understand what computers were.

His choice was considered unusual as most Malaysian students chose to study in the UK for their tertiary education. The main problem with this option was the cost. Tuition fees at the University of Alberta were a mere CDN$550 per semester, compared with thousands of pounds sterling in the UK. Ever resourceful, Danny ventured to the Canadian embassy in Malaysia's capital Kuala Lumpur to learn more about the country. There, he learned Alberta was Canada's richest province and that the University of Alberta was provincially funded. This sealed the deal. Danny knew it was important for a university to be well funded and well equipped so this seemed the perfect choice, as it was cheap and good. The only obvious downside for the boy from the tropics was the weather.

Life in the extreme cold of the Canadian prairies forced him to become resilient. Danny's father had provided just enough monthly allowance, with little left for luxuries or indulgences. For almost four years, he survived on CDN$200 a month, making ends meet by working during the summers as a store attendant in Vancouver's Chinatown and as a fruit picker in the Okanagan region in British Columbia. Upon graduation, he enrolled for his MBA and was deep into his first semester when he sensed something was amiss. While his parents never admitted any struggle, Danny knew his younger twin brothers had just started their studies at the University of Queensland in Australia and that the financial burden must be weighing down upon them. Although he was excited about the job prospects that an MBA would open up for him, he decided to abandon his studies and start working to ease their pressure. Upon returning to Malaysia, his senses proved right; he found out that his parents had to borrow 26,000 Ringgit (about US$11,000 at that time) from his uncle to fund his and his brothers' tertiary education.

Singapore – a city of many attractions to young urban Malaysian youths – was a great destination to start their career. Located on the rooftops of renovated shophouses from the early 1900s are many open-air bars catering to young urban professionals. With a stunning view of the city, many come after work to lounge on sleek banquettes, swallow premium cocktails mixed by skilled bartenders, and to mingle in the electric atmosphere of being upwardly mobile and unattached. With money to burn and few obligations, socializing and meeting other beautiful people is part of the draw of moving to a big city. With a new job as a systems engineer at the Singapore office of global technology firm NCR,[1] Danny would have fit right in with the city's other new transplants, but he was fiercely focused on assisting his parents in providing for his brothers and spent the first few years of his working life 'penniless'. Over the next three years, he gave up personal luxuries so he could contribute to his two younger brothers' education. He did not participate in Friday nights out with colleagues, and was the only software engineer at NCR that did not own a car, using a hand-me-down motorcycle instead. Looking back, his sacrifices paid off when his brothers grew into top positions as senior directors at global giants Pfizer and the *New York Times*.

Three years into the job at NCR, he chanced upon a *Straits Times* newspaper advertisement. Digital Equipment Corporation (DEC), the world's second largest computer company at the time, was looking for regional software systems specialists. Danny saw this as his opportunity to gain more exposure and experience working in different environments. He got the job and began his immersion into the world of a high-flying MNC worker spending considerable time traveling around the APAC region.

It was during his tenure at DEC that Danny was exposed to how technology was changing day-to-day interactions. One project involved the Hong Kong Jockey Club whose system in the mid-1980s could process up to 6,000 online, real-time and off-site betting transactions during races. The Club was DEC's largest non-military installation worldwide with over 250 VAX super-mini and mainframe computers in a single data center. Another project involved email that ran on DEC's machines. Used as a popular all-in-one automation product, it was adopted by many of Singapore's government ministries in their fierce focus on maximizing productivity during the mid-1980s. Perhaps the most significant project he worked on was the world's first electronic road pricing (ERP) project. Danny found himself as a core member of the network design engineering team working with the Philips Engineering consortium. The pressure was palpable with the social cost of international embarrassment to Singapore being at stake. Dealing with high-pressure environments became second nature.

By 1992, Danny had been with DEC for more than seven years and he found himself being head-hunted for the first time. He decided to accept an offer to join Hewlett-Packard's Scientific Instruments business division. Covering the Asia Pacific region for a big MNC gave him both a steady paycheck and a challenging work life with his responsibilities covering all aspects of the product line from R&D to product design, marketing, pricing, support, and distribution. But the biggest change of all was the steep increase in global travel.

Most people think the life of a high-flying global executive – flying from Pakistan, to Guam, to South Korea – is glamorous, but it too often becomes a mental burden. Constantly adapting to strange food, extreme weather, and varying sanitary conditions exhausted Danny and plagued him with a constant sense of being unsettled. He reached his breaking point one day after four years in the job, soon after arriving home after a six-hour flight from Sydney. Although he was mentally and physically exhausted, he could not have a meal and relax in front of the television. His bags needed to be quickly repacked for an evening flight out to Frankfurt for a worldwide product group meeting. He had previously tried to move into a different role within HP, one with less travel, but hit a glass ceiling, as most top-level positions in those days were reserved for US expatriates from corporate HQ. The mental stress of feeling out of control weighed down on his spirit and in that moment he knew his career as an MNC professional might have to end. Armed with a natural aura of confidence, leadership positions have been always been *de rigueur* for Danny Chng. From his days as a school prefect to providing financial assistance for his siblings' tertiary education, his eventual switch from a high-flying position at an MNC to entrepreneurship seems almost a natural extension of his strengths. He decided to strike out on his own and take a leap of faith. It was 1996, at the dawn of the internet boom.

Danny did some soul-searching and decided to pursue the field he knew best: the retail market. Working at NCR had given him an insider's view of retail chains and the problems with the current generation of software. From his days programming point-of-sales (POS) systems for department stores CK Tang and Isetan, it was very clear to him that Singaporeans like to shop. Not only that, but Singapore was, and still is today, a shopping haven for folks from all over the region. In the past, Singaporean fashion aficionados would fly to Paris, the fashion capital of the world, to spend their hard-earned dollars at celebrated fashion houses and niche ateliers. From boutiques in Le Marais to flagships down the Champs Élysées, they joined travelers from all over the world that flocked to the shopping mecca. Today, many are choosing to spend on home turf with a Paris-based market research consulting company naming their very own Orchard Road as the world's best place to shop.[2] But where large technology companies like NCR offered good solutions for such big players, there was a market need for solutions for growing small businesses that were only using cash registers. The decision to focus on bricks and mortar business during the dotcom boom made

Danny the butt of ridicule for some of his peers who thought him old fashioned. But he stuck to his beliefs understanding that:

> Web pages are sexy and nice to have, but mouse clicks cannot be faster than laser barcode scanners. And on a big sale day, which can be often throughout the year, when there are 20 pairs of eyes staring at the cashier, speed matters.

He named his business Sinetics Associates Pte Ltd, rented a small office just outside the central business district and, together with his only employee, a programmer, worked to build his product. To get things off the ground, Danny took out a second mortgage of S$200,000 and put the entire sum into the paid-up capital of the company. While the risk was high, he knew this would give him tremendous credibility when new potential clients did their financial background checks on his company.

During his first few months, he was not able to sign on any customers until several of his ex-colleagues and equipment suppliers sent referrals his way. To seek out more customers, he also asked former classmates, friends, and social contacts to keep him informed of any potential leads, which eventually led to small subcontract work. Then at the end of his first fiscal year of operation, Sinetics was able to gain some momentum by securing two large contracts. The first was for a Southern Rubber Works Sdn Bhd, responsible for popular canvas shoe brand Pallas, which was looking for a turnkey end-to-end retail management system. The second was for Nature's Farm, a Singaporean health supplement retailer. After successfully executing these contracts, Danny found himself playing in the same league as NCR's POS solution space.

As an early bird, Danny caught the worm as a pioneer in the field of POS solutions for small retailers. One example of his pioneering work was the use of hand-held laser-scanning computers for quick and accurate stock counting. Over the next few years, Sinetics grew steadily. He doubled his revenue in year two to more than S$850,000, and held steady in year three, but then growth became stagnant in part due to the after-effects of the Asian financial crisis in 1997. As Danny was the only person performing the role of business development as well as running ongoing projects delivery, he soon ran out of bandwidth for the sales role as ongoing project management required intense attention. Hiring sales executives did not help. The company went through two different sales persons within the span of one year, with both hires unable to deliver any orders. Danny grew anxious. Wanting to pay back his bank mortgage debt, he was determined to find a way to grow his team and sustain the business without borrowing further funds.

In late 1999, Danny gambled on a promising opportunity. While participating in a trade show, the director of systems automation for Tokheim Corporation, a US-based firm that specialized in manufacturing and servicing electronic and mechanical petroleum dispensing systems, stopped by the Sinetics booth. Tokheim was looking for a worldwide solution supplier for their emerging market product offerings in Africa, Asia, Latin America, and the Middle East, and wanted

Sinetics to adapt their end-to-end retail management solution for an integrated POS retail automation system specifically for the petrol industry. They wanted a POS system that could capture dry stock sales, track inventory of the underground fuel tanks, control the forecourt pumps, and integrate everything into one single solution. With an estimated 37 per cent share of the global petroleum dispenser business and a span that covered 80 countries, Tokheim would be Sinectics' biggest client yet.[3] Seeing it as his first big break, Danny agreed to terms that Tokheim would pay upfront upon order and contribute a mere 10 per cent towards development costs. In return for bearing the brunt of development costs, Sinetics would have joint ownership over IP rights. The entire team, which had grown to ten software engineers in all, was assigned to the project. For over one year, Danny and his team focused on development, trial tests, and proof-of-concept runs.

Between 2001 and 2002, Sinetics worked with Tokheim to install 40 pilot installation sites across the globe, from Mongolia to Venezuela. Then Tokheim formerly announced the product called HARMONY, and put it into their price book for worldwide sale. Unfortunately, three months into the product launch in 2003, Tokheim filed for Chapter 11 – a form of bankruptcy in the USA that allows corporations time to restructure debts. A short time later, the SARS epidemic hit the region, halting business travel to many Asia Pacific countries, and having a devastating effect on Singapore's economy. In an instant, Sinetics' financial lifeline dried up. Since they had not attended to any other leads when they decided to focus on this project, and left without any other alternative, Danny began to look for VC funding as further bank borrowing was not an option Danny wanted to explore.

An old friend from high school introduced him to iSpring Capital, one of the several specialist VC entities appointed by the Malaysian Government Venture Capital Fund (called MavCap) from Malaysia's growing VC scene in the early 2000s. In 2001, iSpring Capital was selected as one of the VC firms to manage a RM$50 million venture capital fund allocated by the government because of their forward-thinking approach to investing in technology start-ups. (Adnan, 2002) In Malaysia's growing entrepreneurship ecosystem, many firms, especially those held by the government, eschewed risk and were hesitant to invest in start-ups. iSpring Capital however saw potential in Sinetics and agreed to invest S$250,000 as phase one seed money to help Sinetics continue their development in the niche petrol station POS automation solution business. The VCs were friendly and all came with finance training and background. However, the firm took an aggressive approach to pulling Sinetics out of financial trouble by insisting Danny chase after many of their leads. They wanted to attract big customers and big projects, and recommended Danny chase almost anything in the petrol station retail sector around the region. From their suggestions, Danny knew they meant well but lacked industry know-how. Unfortunately, he felt obligated and he refrained from arguing with the VCs.

Holding 33 per cent of the shares, they had a strong seat on the board and Danny needed their financial support so he tried it their way for one year. To his utter dismay, the majority of their leads did not materialize into new projects and Sinetics was quickly running out of money again. The little revenue they had was from maintenance support from existing retail management solution customers. At this point, the company had no other retail customers as Danny was advised by his VCs to drop his focus in the 'generic' area and concentrate in the niche vertical 'oil and gas' retail automation market.

The stress took its toll on Danny and his health and personal life began to deteriorate. He was faced with a sizable bank debt due to the negative equity of his office unit property and an ongoing divorce proceeding; he also underwent an angioplasty operation due to a blocked artery. He had finally hit rock bottom. Even though, by that time, the VCs were in the midst of preparing for a second round of funding in order to try and save Sinetics, Danny saw no point in continuing to fight when there was no value in the leads provided and no synergy from the VCs.

He decided to call it a day and scaled down (and eventually closed out) the entire software product development team, and by this time, the VCs also gave up and agreed to exit their stake in the company for S$1. It was also during this time that he decided that he had to make good after nursing his health back. Highly sought after because of his diverse arsenal of technical and management experience, Danny quickly found himself being head-hunted back into the MNC sector for a series of senior executive-level positions. It was an easy transition back to being an employee for Danny, as he was relieved of the pressure to pay salaries every month and to balance a 'shoestring' budget.

Despite a difficult first attempt at entrepreneurship, Danny does not rule out trying it again, especially now that he is debt-free after repaying all his bank debts. Perhaps had he stuck to his fundamentals and stayed conservative in pursuing and maintaining organic growth in the bricks-and-mortar business of traditional retail management solution market, and not gone into the petrol station niche, he might have already made enough for a decent and comfortable retirement. But then again, had Tokheim not gone into Chapter 11, he might have made millions with a niche product that would have been usable worldwide. But, he says, that is a lot of 'ifs'. Danny has always been an ideas guy, a ringleader, and ran into difficulty when he was unable to articulate himself, and therefore was unable to hold his ground with his VC shareholders. Admitting his shortcomings, he reflects that next time around 'I would not make the same mistakes. I would also have a strong operating partner to share the load and, most of all, to provide sanity checks to every major decision affecting the entity'. Leaving the security of a well-paying job is a chance risk-takers take in exchange for sky's-the-limit growth potential. Both options have their pros and cons. At the time he left HP, many of his colleagues thought he was crazy to spin out on his own. Sometimes, these workers who choose stability over quick growth are locked into a salary category, but some of his peers in the HP printer group ended up wealthy through a steady income and judicial investments. Of those that did spin out, the most successful ones focused on becoming suppliers and contractors for the parent company since they already knew what was required. They did not choose the path of creating a new technology product company with development costs and uncertain outcomes.

From a small town in Malaysia, to the flat lands of Alberta, to the bright lights of Singapore, Danny perseveres by observing and learning from his environments. His experiences have cemented in him the importance of communication, setting the correct level of expectations and dialogue with investors. Sinetics' investor had good intentions but no experience in the technical field; this is inherently true of the majority of VCs who are often fund managers with financial backgrounds. Building a respectful relationship can be difficult when there are competing interests, but the most successful start-ups are those where the entrepreneur works with the investors but the VCs also work to make sure the company gets the support it needs. When given the chance, most VCs intervene and give their opinion on how the company should be run, but it is the CEO's job to decide on whether he takes their direction and to communicate his reasons for taking or not taking actions.

Lessons learned and key takeaways

Accepting capital from the right investors is a key to success

A common threat in all three cases is the unmet expectations of the VCs post-investment and the subsequent miscommunication between the founder and the investor group. In Star+Globe and Pixelmetrix, both companies raised money right before the dotcom bubble, and their investors wanted them to pursue a quick-win, 'dotcom' positioning so to enable a fast exit. Both companies were building deep technology products – selling B2B, which necessitated a long sales cycle – so this expectation could not be easily met with merely a strategy pivot. Entrepreneurs must take money from the right investors who share their vision of building a technology company and who also understand their business. In Star+Globe's case, raising too much capital resulted in a power shift from the entrepreneurs to the investors and the excess cash became a reverse merger target. Pixelmetrix was able to manage this mismatched expectation by learning to communicate effectively with their board. However, Sinetics was unable to resist misguided investor directions and paid the consequences. All three companies had to expend resources that were not productive to the original product vision due to the need to placate their investors. Taking money has a cost. Technology entrepreneurs should raise money with this knowledge in mind.

The three narratives speak to an important set of lessons:

1. Do not run out of money.
   Technology entrepreneurship is a long journey, seldom are there instant successes. You are not the next Instagram. Young companies need to have the cash cushion to withstand the uncertainties and external shocks, which often derail a company. Sinetics ran out of money from a combination of bad luck (principal client going Chapter 11), external forces (Asian financial crisis and SARS), and inappropriate direction-setting by the investors.
2. Raise money from the right investors.
   Most investors are from banking or finance background. Very few have the contextual knowledge of what it takes to build a winning technology product. It must be understood that investors come with a different perspective so differences in the way forward are to be expected. Star + Globe did not

anticipate the lack of technology understanding from their lead investor and did not manage the differences effectively.

3. Realize that taking money comes with a cost.
   When entrepreneurs accept capital for equity, the company becomes shared with external parties. You cannot ignore your co-owners. Star + Globe did not expend the attention necessary to educate the investors on its future direction and intended strategy in the e-business transformation. A more united board with deep support of the articulated strategy would have stood a better chance to vote against the disastrous merger.
4. Effectively communicate with your investor post-investment.
   Effective communication builds mutual trust and prevents investors from making counterproductive interventions. Although Pixelmetrix was able to stand firm on pursuing the technology leadership vision, the company did not make use of the investors' network to facilitate its growth. Sinetics' investors gave directions that did not productively create sales results for the company.

Part II

Plunging into the market

Managing dilemmas and keeping afloat

The stories in Part I chronicled how three technopreneurs managed investor expectations and the importance of communication. In Part II, the journeys of four technopreneurs who started and grew their ventures with bootstrap and customer revenue will be shared. In three of the cases, the technopreneurs started a solutions integration venture with the product development efforts either funded by a series of customer-funded refinements or by merging with product companies. All four technopreneurs found profitable exits from a buy-out offer from customers or large corporate companies needing access to the venture's technology know-how.

4 Finesse Alliance and Chak Kong Soon

Streets in Malaysia, whether it is Jalan Bukit Bintang in Kuala Lumpur or Jonker Street in Melaka, testify to how traditional elements successfully buoy themselves in a sea of modernity. Even with social media claiming the free time of Malaysian youth who tweet their friends and check Facebook while walking, they remember to nod their heads politely when passing an 'auntie' or 'uncle' in the street. Deeply ingrained in Malaysian culture is a practised deference for elders and a strong respect for family. Families remain close-knit with children staying connected to and sometimes living at home, even after they get married. Even the most modern urbanites with a taste for big city life accept a move to his or her fiancé's hometown after marriage in order to stay close to the family hub. Cohesiveness is important to Malaysians, with traditional values such as respect, cooperation, and loyalty greasing the wheels that are driving their rise as one of southeast Asia's 'ones to watch'. In this culture, many moves in the game of life are made after consulting with various family members, although a growing number of trailblazers are managing to carve their own path. An increasing number of young Malaysian's continue to revere their elders while branching out to pursue new territories.

Thirty years ago, Chak Kong Soon was wrestling with this situation. His parents, both trained teachers, wanted him to earn good grades and segue into a comfortable middle class lifestyle. A few years earlier, his father had done some work for the Indonesia's Ministry of Revenue or Taxation (known as EDIRC) and afterwards set up his own lucrative accounting practice. Their dream for Chak, their eldest son, was to take over the new family business. Chak definitely had the talent and character to succeed at the task set before him. While he did not score at the top of his class, he was a good student and a strong student leader. As an 18-year-old junior college student, he served as his school's head prefect (known as 'school captain') for both the boarding house and the school, and as an athletic captain for the track team. Popular and supported by his teachers and peers alike, his secret to success was a likeable and hardworking personality that easily convinced others to plug into his vision. With a strong and determined spirit, he had a long history of successfully accomplishing whatever he put his mind to. Unfortunately for his parents, he did not want to pursue accounting, no matter how strong his reverence for them was. He was studying computer science at the time and knew the IT industry was exploding in the region. Instinct told him computers were the next big thing and he wanted to be a part of the growth despite not enjoying his junior college courses in the subject. Chasing the future was more exciting than the stability of an accounting degree. Understanding their son was a bit of a trail blazer, Chak's parents relented to send him abroad for university studies.

At the time, studying abroad was a coveted opportunity for many middle class Malaysian students who aspired to improve their English skills and other soft skills needed to succeed in the global marketplace. The hip cities of London, Melbourne, and New York have long been popular destinations to study, but Chak surprised his family with his choice. Chak chose his dream school not based on the city's architecture, sandy beaches, or production of popular culture but for the niche course offerings. At the time, most universities only offered courses in computer science or business applications, but Chak wanted to study management information science. He found his dream program at the University of Iowa and flew off for the rolling green hills and brown corn fields of Iowa city in 1984. Over the course of his studies, he immersed himself in understanding how to architect database applications, create business solutions, and develop troubleshooting skills for infrastructure problems, and graduated two years later with a bachelor of business administration, double majoring in business applications and management information science, with a minor in computer science.

Ready to make his mark on the world, Chak took a look at the global IT industry and decided to go where the money was. When he graduated in 1986, the IT industry in Asia was hot. Drawn to the bright lights of emerging metropolis Singapore, he moved to take a position with 'Big Five' accounting firm Arthur Andersen (the former parent company of consulting giant Accenture) as a programmer under their Management Information Consulting Division. At the time, Andersen was renowned for transparency and honest business practices. A position at a reputable global MNC was a dream job for many graduates but Chak's mother had her concerns. Like many in her generation, she did not know much about the firm or its reputation and was concerned Chak was not reaching his full potential. Her fears were extinguished over the next few years as she saw her hard-working son ascend ranks exceeding her expectations of a comfortable middle class lifestyle for him, moving from programmer, to analyst, to joining the standard consulting practice. Within four years, Chak had obtained a managerial position achieving it faster than his counterparts by two full years.

His superiors noticed his leadership and sent him on an overseas assignment to the African island of Mauritius. Blessed with powdery white sands, lush sugarcane fields, and surrounded by turquoise waters, Mauritius was a dream work destination. There, as a senior manager, Chak was running the IT department for a local bank managing eight Andersen personnel and around 40 client personnel. Officially he reported to the bank's head of IT but essentially he ran the show experiencing a relatively high degree of autonomy. Working for an MNC from a beautiful outpost was not a bad change from slogging through Singapore's competitive urban jungle, but he soon grew restless. Overseas assignments were exciting but he knew such moves would become complicated once he started having a family of his own. He highly prized family life. He had heard stories of one co-worker who had been promoted to lead an overseas office but the assignment had unintended consequences:

> One partner at Andersen flew the family there and the air wasn't good. The kids got asthma and he had to get them back and forth [to Singapore], and this happened for four to five years. That's something I did not want to go through.

The main reason to stay at Andersen was the potential of making partner and Chak was already on the right path and being groomed for more responsibility. With this aspiration, he knew he would have no choice but to go if he were asked to lead another overseas assignment. There was also the option of jumping ship to a number of banks that had contacted him about becoming their assistant CIO, but working at another MNC, even at a position with more responsibility, did not appeal to him. The increased financial compensation and sophisticated work tasks were tempting, but he began to accept that partners still have to toe the line no matter how high they rise. To the surprise of his co-workers, he decided to leave in 1995 after nine years with the firm to start his own business. At age 30, he felt he had little to lose. With his track record, he could always get another MNC position if he failed at entrepreneurship. After making the decision to spin out, he hired a secretary and a young technical whiz, and began to lay the foundation for his new venture.

What was a young maverick to pursue? Out of his nine years at Andersen, seven years were spent in the financial services division so he decided to leverage his expertise and provide software solutions for financial institutions looking to build electronic and mobile commerce capabilities. He quickly realized that his Andersen experience was the perfect foundation for starting his own business: 'Andersen actually provided me with the closest entrepreneurial experience that I could get because once you became partner of the firm, you needed to run the firm'.

From day one, he decided to build his business like the MNC he spun out from even on a start-up basis with proper accounting, processes and so on. This way, he thought, there would be an objective-driven methodology where staff could make their own decisions and still fall in line with the direction the company was going. There would be an infrastructure and methodological way of getting things done. Dipping into S$100,000 of his own savings, he started Finesse Software that September.

Finesse specialized in making finance software, namely for system integration in banking and financial services. Within two months, he signed on his first client introduced to him by his social network consisting of ex-Andersen colleagues, friends, and connections at the banks that had wanted to hire him. Finesse grew quickly aided by a flourishing industry, but the growth halted after two years with the onset of the Asian financial crisis. With luck on their side, they managed to complete and financially exit from projects in Thailand and Indonesia allowing them to collect their revenues during the calm before the storm. But the storm continued to brew with Chak's contracts conflicting with those of his former employer's for the first time. Chak was under no legal obligation, but Finesse's presence became an irritant for Andersen as they sought the same clients. Chak wanted to keep the relationship clean and did not approach banks he worked with under Andersen to get jobs but they began to overlap in another way. Usually their projects would not overlap, but in the aftermath of the crisis, Finesse saw their projects moving from the S$500,000 to S$2 million range while Andersen saw their projects come down from the S$5 million to S$2 million mark. Although Andersen had their reputation, their large size meant processes often took longer than Finesse whose small size made them more agile. Rather than build organically through a hire and fire policy, Chak's strategy for growing Finesse was to join up with partners. From his days as a student leader to his time as an MNC worker, Chak's peers always branded him as fair. His natural inclination was to support others and work together to achieve success. A charismatic leader, Chak, as CEO of Finesse, attracted many interested partners and the next few years became a rollercoaster ride of mergers and acquisitions.

A few months later, he got some projects working in partnership with another ex-Andersen employee named William Wong and, after four to five months of cooperation, and to maximize their market impact, merged with his company, WinSoft. The two owners were close and ran the two companies separately but in tandem. Chak continued to work on the technological and consulting side of the business, with William taking on managerial tasks such as head-hunting and software sales. At this point, the companies had about 35 employees in total. Over the next three years, the fruits of their partnership continued to blossom and they further spread their influence by acquiring a software company, which was a Lotus Notes developer, a collaborative client-server application. With this third part of the triumvirate, they targeted themselves as the top Lotus Notes developer in the region for financial services. Then, after two and a half years and exceeding many goals, they realized that the three businesses were not sufficient for their constant growth and looked for a software brand they could buy. Their goal was to scale up, to become a software developer and owner of IP. One of the companies they were working with to develop software, and who they knew from their working relationship with Standard Chartered Bank, was seen as such an asset – and so in 1997 they merged with Elcom Software. Having grown to 70 employees and continuing to accelerate at bullet-speed growth, Chak and William decided to merge all three companies in 1999 to become Finesse Alliance. With these actions, Chak was practising both effectuation and goal-seeking actions (see boxed text on the contrast between effectuation and causation).

With three mergers under his belt, it was time for Chak, now as CEO, to seek VC funding. In doing so, he cautioned that that they only accept funding from a VC in tune with Finesse's DNA. Finesse found their ideal match with Hong Kong-based Crimson Ventures. Funded by bank owners, the partners understood the nuances of Finesse's software and were able to open a number of doors for them.

Effectuation versus causation

Effectuation is often depicted as a contrasting logic to causation (Sarasvathy, 2008). According to effectuation scholars, the two contrasting logics manifest in the following manner:

	Effectual logic	*Causal logic*
Beliefs about the future	The future is to be enacted; there is no fixed future. Key to creating the future is deployment of means.	The future is predictable. Strategies can be devised to take advantage of predictions of market preferences and trends.
Actions	Means drive the goal determination. Means are: what I know, who I am, and who I know. Goals emerge from the effects after actions. Allow for *experimentation* on how you can achieve new effects from means.	Goals determine what actions to take – including necessary steps to assemble resources to achieve goals.
Unexpected events	Surprise can be a positive leverage. Be *flexible* on how unexpected events can open the future for opportunity.	Avoid surprises. Focus is on prediction and planning.
Risk	Instead of investment perspective, risk only what you can *afford to lose.*	Classic trade-off decision on risk and expected return on investment.
Third parties	Partnering viewpoint. Focus on *partnering* with others and *pre-commitments* to create returns.	Competitive viewpoint. Maximize your ROI.

Source: Sarasvathy and Dew (2005, Table 1, p. 390); Read et al. (2009, Table 1, p. 3).

As can be seen from the narratives, many of the technopreneurs practised both forms of logic as they evaluated opportunities and strategies for navigating in the market. Many started their new ventures as the smaller, more nimble versions of their previous employers. They relied on their means to generate the first business opportunities, even as they pursued their goals of growing the business. Chak, in particular, was a prime example on use of both causal and effectual logics. Chak knew after a few years in the market that he needed to evolve Finesse into a product company to maximize the value of the company. He set out to achieve this goal by partnering with the few companies he was already working with or knew of from his contacts, and created a merged entity and refocused the solutions side of the business to align to the product focus of Elcom.

There was also a strong degree of trust in their relationship, and Crimson never interfered with Finesse's operations. From his own experiences and from knowing about others in the field, the importance of selecting the right VC cannot be overemphasized. Most start with angels and move to VCs when they need serious aid. Chak emphasizes that this decision should not be taken lightly as entrepreneurs must seriously consider how much of the company they are willing to give up. For example, some VCs only invest S$5 million and above, and should owners of a company worth S$10 million want to accept their funding, they must be prepared to give up 50 per cent of their company. Also important is to look at where the VC's are in their fund cycle, the danger being that a VC might be ready to close their fund, when a company feels ready to IPO.

Additionally, VCs can open many doors for budding start-ups so it is important to consider finding a VC with connections that match their interests, whether they be in government, corporations, or in a niche industry. Looking back, Chak says, 'The biggest issue with the dotcom or with the rush in getting capital was that a lot of entrepreneurs got stuck with VCs and did not know how to exit from VCs'.

In 2003, Chak and the team sold Finesse Alliance to a partner in Malaysia that saw their business as a way for them to build upon their Islamic banking products. He reflects that not being of the religion would hinder their ability to sell to the Islamic market, although they understood the software. Many businesses agree to be acquired for a handsome payout but Finesse and Crimson did not make financial exits. Instead, they did a share swap – they all received shares of the new entity instead.

Returning to the MNC industry, his first position saw him spending weekdays in Hong Kong running and building the services business for Vanda, a hardware distributor. Under famed Hutchinson Whampoa's wing, Vanda is listed on the Hang Seng Index, and working there was a great eye-opener for Chak. He met many high-profile people apart from Li Ka-Shing himself. Li Ka-Shing is the chairperson of Hutchison Whampoa Limited and is currently he is ranked ninth on Forbes list of billionaires. Despite excelling at his job and being invited to stay on, Chak decided not to renew his contract. He did not want to move his family over from Singapore, and climbing the corporate ladder at 40 did not appeal to him. After thinking over his options, he realized his passion for entrepreneurship still burned brightly. He says 'I find thrill in making things happen even though things [should not be able to] happen, and that's where the spirit of entrepreneurism comes in'.

In 2004, he officially submitted his resignation and decided to venture into consulting. Having successfully built Finesse in Singapore when VC funding was in its primary stages, he had a great store of insight to share with the new generation of budding Singaporean entrepreneurs. He was seeing many companies achieving decent levels of success, but had lukewarm relationships with VCs that were not happy with the returns on their investment. There were also a number of external factors affecting start-ups at that time. The Asian Financial Crisis had weakened the market, but the dotcom crash of 2001 had really taken the industry for a beating. Then the SARS crisis further deteriorated investments in the region. Using his arsenal of knowledge, Chak advised start-ups and mature start-ups on how to use VCs and how to exit. In consulting, he found a happy niche area where the work evolves, constantly building a fast-paced industry where he is happy. Through consulting, his eyes were opened to another area. He was noticing a pool of high net worth individuals that were not content with planting shrubs, growing trees, and taking care of their grandchildren post-retirement. This group was already making ends meet financially and wanted to use their money to build new things. What they craved was putting their skills to use, but knew they did not have the energy to pursue a start-up of their own.

Chak moved to fill his gap by starting an investment firm called Stream Global in 2007. Along with friends Wilson Tan and Bill Liu, they designed the firm to be a platform to encapsulate their vast store of business and IT industry experiences and pass them onto the new generation. With extensive experience in both the corporate and entrepreneurship scene, they all brought their rolodex of contacts. However, Stream quickly turned from an advisory to an investment vehicle. Fresh ideas inspired the team, which wanted to see them through to the next level. They began to talk with the NRF and declared it was not enough to simply give out money. Budding companies needed real help; this likely fertilized the soil from which the Technology Incubation Scheme (TIS) sprouted. More than anything, the team saw themselves as mentors and consultants for start-ups; a structured group of angels in private equity more than a traditional VC firm.

Many in Singapore's start-up scene had been paying attention to Chak's journey. Amazingly, he had led Finesse through the Asian financial crisis, the dotcom crash, and the effects of the SARS epidemic on investment in southeast Asia. He had also exited from a VC relationship unscathed. Many wondered just how he did it and Chak, always wanting to help others achieve their goals, started consulting for start-ups, eventually gaining enough clients to make it his full-time job. What worked for Chak was inorganic growth. He took a chance and pursued opportunities as they arose. While doing so, he stuck to his true north and was very selective, rarely allowing himself to be swayed by the opinions of others. 'Be yourself, nurture relationships, take care of your family, and respect others', he encourages young entrepreneurs today; wise words from one of Singapore's most seasoned entrepreneurs and mentors.

5 PlaNET and Ronnie Wee

From a young age, Ronnie Wee aspired to be a doctor. His father was a cardiologist in private practice at Mount Elizabeth Hospital and Ronnie often followed his father around on his hospital and house calls. His father had high expectations for his only son. In Singapore's competitive ecosystem, tiger parents who wanted the best for their children often pushed them into the respected and coveted position of doctor; the more difficult the specialization, the better. Comforting, healing, and saving the sick is an heroic task granted to the most disciplined students and Ronnie was up for the task. At the Anglo-Chinese Junior School, he stood out as a well-rounded student and high achiever. He was on the tennis team, while also playing football, table tennis, badminton, and squash, and was on the prefectorial board. Accomplishments aside, what made him most popular among his peers was a congenial yet mischievous nature that belied his unassuming stature. Reflecting back on his time as prefect, he jokes 'The teachers had to nominate me and I was also voted in by the rest of the school. I think it was a combination of leadership and also my being a small guy ... everyone thought I would not bully them!' His parents, confident in his capabilities, sent him to the UK to study for his O-levels, initially as a precursor to getting into medical school in the UK. There he earned 11 O-levels including Chinese, French, and German.

Ronnie was one of the lucky ones. In Singapore, parents sacrifice to give their children the best education they can afford with international stints in England and the USA being most coveted. At the age of 14, Ronnie flew to Hertfordshire, England to board at Haileybury College, where he was on the school tennis and squash teams. He achieved his school colors for both and was the youngest (and smallest) member ever of the tennis team. He had to fend for himself in the military-based school and being on both sports teams earned him respect from his fellow students and teachers. He also won academic awards in English language and English literature. His O-level results, combined with his extra-curricular activities, were strong enough to earn him a place at Tufts University in Boston at the early age of 16. At the time, the Singapore Government had a policy that if a student could get into university before the age of 17, he could defer national service[1] until after graduation. It seemed a good idea at the time although one would argue 16 is a tad young for the 'best years of your life'. Still, it would continually challenge Ronnie to be immersed in an environment where he had to grow up very quickly.

In the 'City on a Hill', Ronnie's eyes were opened to a liberal lifestyle vastly different from the more controlled and disciplined environments he was used to. After spending his freshman year as a pre-med student, he admitted to himself that medicine was not his true passion. He craved action over the delayed gratification of studying to be a doctor. At the time, the IT industry was getting hotter so he tapped his network, sought out some of his engineering friends, and realized that their degree allowed them to work right after graduation. That sealed the deal for Ronnie. He began the slow process of transferring to chemical engineering, a safe bet because these grads had the highest acceptance rate into medical school. To further explore his engineering interests, Ronnie became involved on campus. He was voted in as president of the Association of Aeronautics and Astronautics, an active member of the Institute of Engineers, and in addition started a printing company with some friends, targeting printing resumes, report,s and brochures for students at Tufts as well as neighboring schools. The profits provided a good source of income for meals, books, and travel.

After graduating with a BSc in electrical engineering in 1986, Ronnie moved back to Singapore, as required by the Singapore government, to undertake his national service. He went through the usual basic military training and then went on to officer cadet school and became an army officer. Initially, national service was a huge culture shock for Ronnie who had just spent the past six years studying overseas. In hindsight, though, he felt that this was an extremely important part of his life, as it enabled him to bond with fellow Singaporeans from literally all walks of life. His time at cadet school taught him important leadership and perseverance skills.

The time came for him to put his degree to good use, but he was faced with a dilemma that plagues so many twentysomethings. Ronnie knew what he was passionate about, but was unsure of how to translate that into a career. He decided to pursue a typical route for smart Singaporean males and applied for positions with global banks and even with Singapore Airlines. He approached a friend who had a laser printer at the time to assist him with printing his resume and that friend happened to be doing an internship with the Management Information Consulting Division of Arthur Andersen. She basically convinced Ronnie to apply to Andersen to be an IT consultant. He had never considered consulting before, but hearing his friend speak of a job that allowed him to cultivate a well-rounded understanding of a client's current problem, derive a solution, build an ongoing relationship with the client, and then advise them on their future problems appealed to Ronnie's inherent knack for problem-solving. The friend passed on Ronnie's resume to the partner-in-charge and an interview was set up. At one of the many interviews, Ronnie remembers the partner-in-charge 'pooh-poohing' the other companies that Ronnie had applied to, saying that they were all Andersen's clients and what better way to work for all of them than to consult for them. Ronnie was sold.

The workplace culture of a powerful organization can be potent. Fresh employees drink in their newfound marketplace prestige taking on the persona of a confident, untouchable MNC worker. At Andersen, rivalry among the new army of fresh consultants was fierce. Each tried out the firm's different faces – learning about program structures, how to tie business needs to technology solutions, and how to unlock the potential and value of a client – and smart employees worked hard to be a standout among the competition in the corporate labyrinth. Ronnie stood out early on by focusing on a technology-based role. The banking industry fascinated him and he wanted to learn about the technology connecting a bank's various branches. In those days, mainframe technology aimed for client-server architecture had to go through a network and the focus was on how to design networks to support the banks' needs. He settled into the network division and consulted on the necessity of basics such as local area networks (LAN), wide area networks (WAN), and the internet. One of Ronnie's first projects was to design, implement, and operate network architecture for one of the largest banks in Indonesia, and to connect its head office with over 200 bank branches throughout the country via a combination of leased lines and satellite technology. It was the first of its kind in Asia.

For the next seven years, Ronnie's career jumped from strength to strength and he was on track to make partner. He got to travel extensively, planning, designing, and building innovative technology solutions for large banks across Asia-Pacific, Europe, and North America. Ronnie led the inception and development of the Technology Integration Services division of Andersen Consulting (AC) and assisted to build over 100 specialized technology consulting organizations spanning southeast Asia. He was promoted multiple times, ultimately becoming a manager in 1994. But after several years, the thrill of starting his own company became increasingly attractive. After all, he had already assisted to build a successful 'company' internally for AC; surely he could do it on his own.

At about the same time, he had been approached by a number of head-hunters scouring AC staff for their technology and consulting competencies, enticing them with two and a half to three times their current salary, but he turned them down. 'Why leave the best technology consulting company in the world unless you were starting your own company' used to be whispered in the hallways. He had once opted against the stability of pursuing a career in medicine; this time he chose to opt out of a stable MNC career. He felt that he could give himself two to three years in the market to see if he could make it work and, if it did not, he could always return to AC and at least 'be a photocopy guy' or something. After close to eight years with AC, Ronnie put in his notice and left the firm in 1996. Everyone was surprised, as he was doing very well and on a progressive career path, but Ronnie was ready for a new challenge.

At that time, there were also a number of consultants leaving AC to set up their own businesses. Just as officers band together as brothers during times of war, Ronnie and fellow Andersen-consultants-turned-entrepreneurs supported each other when they spun out. They understood each other's dilemma. While working at a big firm was challenging and rewarding, this group struggled with the fact that their hard work was building someone else's organization. Excelling in the intense atmosphere perfected their presentation, consulting, project management, and technology solutioning skills – Ronnie believes eight years at AC was likely the equivalent of working 10 to 15 years at a competing firm. Giving up a shot at partnership was a struggle they all faced but, as senior managers, they knew the time was now or never. The opportunity cost would be too great if they waited any longer. One entrepreneur, Chak Kong Soon (profiled in the previous chapter), asked Ronnie to join him and his company in a co-op space where they would share resources. This arrangement was ideal for Ronnie as he had left his job with only a vague idea of what he wanted to focus on.

Working at AC showed him an untapped industry niche in consulting for smaller businesses on how they could introduce and leverage technology to achieve their business goals and objectives. With an inbred inclination to help others and assist the underprivileged, Ronnie sought to carve out a plan for a technology consulting company. That August, he also welcomed the birth of his daughter, Camille – so he had effectively two babies that he was incubating. He named the company PlaNET, choosing the name because he was very impressed with the work that Sun Microsystems was doing at the time with networking, Java and the internet. PlaNET hence revolved around the SUN and built many of its early solutions on the Java platform. Carrying himself with the assured confidence of a professional on a high from choosing to leave a promising career at an MNC, he expected a hearty welcome from large corporate clients. He quickly realized that as great as his work and reputation were, clients wanted the AC name as an assurance of quality. Reflecting on this, he says:

> One of the rude shocks I had after coming out from the corporate environment is that as much as the clients 'loved' us and the quality of work that we did, it was really the Andersen Consulting brand that they were after and not the individuals' work. We were turned away from many clients even though our fees were a fraction of what AC charged. It then also dawned on me why corporations really hired consultants. It was so they could have someone to blame when things went wrong. They could sue Andersen Consulting but they could not sue PlaNET.

Lacing up his bootstraps, Ronnie worked to build PlaNET from the ground up through a fee-based consulting model. He hired some consulting and IT people, mostly fresh graduates and those with less than three years of experience, and focused on providing services such as web design, e-commerce applications, and turning solutions into products. When they first started, they did not need investments. As a systems integration consulting division, the initial injection of S$100,000 from Ronnie and his father was enough to fuel the company, which became cash positive after three months. Over time they branched into network and infrastructure consulting for high profile clients such as Caltex, GE, and global banks. Despite this success, PlaNET continued to struggle as their model was not scalable. The challenge was that Ronnie's clients refused to have anyone but him working on their projects, and he began to feel like an MNC professional again. At this time, he went to talk to a few investors but most of them wanted to wait six to 12 months. It was during the dotcom boom and they were looking for revenue and wild ideas and the multiples they were giving were very low. With his background in banking, Ronnie looked for a way to scale the business and found his solution in internet-based technologies for banks and brokerage firms.

The team designed an initial suite of products around a financial portfolio management solution using internet technology and then recruited a few SME clients and Japanese investment funds to use them. The product was called AIMS – Advanced Investment Management Solution. AIMS enabled both small and large investment firms, and individuals, to manage their financial portfolios in real time over the internet. It was a combination of Ronnie's consultative selling skills and one of the first internet-based portfolio solutions that made it very interesting for the early clients. Basically, companies no longer needed to spend large amounts of money on software licence fees and instead could share and view investment data securely over an intranet and the internet. Also the financial data was live since PlaNET had built an interface into Reuters and S&P which provided the live data. This had never been done before over the internet. A big break came in the form of a significant order from Dai-ichi Securities, which used it to share investment data between all their offices in Singapore and in Japan. Charged by this success, they approached large brokers such as DBS Securities, Phillip Securities, GK Goh, DMG Securities, Kim Eng, and Citibank's internet banking and internet trading divisions. DBS Securities and Phillips Securities were very interested but eventually decided to build the solution on their own, while DMG and Kim Eng signed up as initial clients. Over time, the quality of PlaNET's solutions became known throughout the industry, with AIMS receiving a nomination from the InfoComm Development Authority (IDA) as one of the most innovative products.

Ronnie's keen eye noticed that the internet was gaining momentum in the western world with some market leaders, such as the Charles Schwab Corporation and eTrade, using it exclusively for trading. Anticipating in its growth potential, he decided to pilot it for the Asian market, believing the trend would soon move eastward. He was laughed at by a few who said the internet would never replace the old way of trading since it depends on relationships, but his risk paid off two years later when PlaNET was granted the opportunity of a lifetime. In 1997, Bill Gates came to Singapore on the invitation of the Singapore government and PlaNET was one of eight companies picked by Microsoft to present him with their internet trading and portfolio management solution, using a then revolutionary mobile Windows CE device used to check stock prices and purchase stocks on the move. Today, that seems commonplace, but back then Bill Gates was pushing very hard for mobile solutions. He was so impressed with PlaNET's solution that he spent almost ten minutes at the PlaNET booth asking a series of questions and giving his suggestions for improvement. PlaNET was also featured on the Microsoft global website as a promising innovative company from Asia.

Meeting with the (at the time) world's richest man was certainly a highlight for Ronnie, and their discussion gave him a lot to ponder. How was he to take PlaNET forward? Their product was becoming more robust, but PlaNET was not really a product company, but nor was it an internet company. Taking stock of their assets, it was logical to project themselves as a solution or consulting company but this made it difficult to attract funding. At the time, however, most VC firms were interested in pure internet companies such as B2B and B2C companies like eBay or Alibaba. Firms like PlaNET were tier 2 on their watch list. PlaNET had then approached a variety of investors – including Citibank Ventures, EDB Investments, Walden, and Warburg Pincus – but they were turned down since most of PlaNET's revenues came from consulting and systems integration and were thus not scalable.

Ronnie then made the decision to focus entirely on turning PlaNET into a product company and stopped all consulting and systems integration work. The company recruited a number of systems integration partners to sell its product in Singapore as well as overseas, eventually managing to close deals with brokerage firms in Singapore, Malaysia, Indonesia, Thailand, and Hong Kong. Throughout 1998 and 1999, the internet was really starting to take off in Asia and companies started to receive funding on crazy valuations. Corporates were scrambling around trying to figure out how to get into the internet game and take advantage of the hype.

One such company was Datacraft Asia Ltd, a Singapore-listed network systems integrator specializing in Cisco solutions across 13 countries in Asia. ING Barings had written an analysis report about Datacraft in early 1999, saying effectively that margins were thinning in the systems integration space and that, for growth, Datacraft needed to focus on moving up the value chain and building a solutions business. Datacraft also wanted to develop an internet strategy for the company and saw the acquisition of PlaNET, along with a number of other Asian internet companies, as a logical growth strategy. Datacraft approached Ronnie towards the end of 1999 and made an offer to buy a majority stake in PlaNET. They also received a few other offers from SMEs, listed companies and VCs alike who all saw PlaNET – with steady growth, good client base, and proven products – as a low-risk investment with a high chance of positive returns.

For Ronnie, it was most important to sell to a company in tune with PlaNET's DNA. He say:

> I thought the synergy was the greatest and strongest in the Datacraft deal. We would bring the know-how, the products and solutions, and they would bring their footprint, sales force, financial infrastructure and client base across 13 countries across Asia.

Even though the valuation paled in comparison to some of the lofty valuations at the time, Ronnie decided to sell, hoping that the unique and powerful combination of innovative product and sales infrastructure would bring PlaNET to the next level of growth. Staying on for three years after the sale to smooth the transition, Ronnie worked to productize and push through the channels.

Fast forward three years, his division grew from making S$1.5 million to S$5 million. Despite this growth, the industry was hit by SARS and 9/11, plummeting Datacraft's shares from S$11 to S$0.50. Desperately trying to save the firm, a new management team came in and brought the company back to its roots selling hardware and systems. Originally, Ronnie's team as well as a number of others had been brought in to build solutions consulting and solutions business but Datacraft's business model was very hardware and systems driven with huge volume but small margins. Datacraft's sales teams were all trained to sell network solutions that included basically hardware and system software. They were not used to the consultative solutions selling approach, which involved a good understanding of the combination of business and technology. Because most of those in Datacraft's sales team were 'box-pushers', and were trained and incentivized as such, there was a significant disconnect between the PlaNET consultative solutions selling approach required for software and services and the hardware-selling approach of Datacraft. Given the market turmoil at the time and the fact that Datacraft's market cap had collapsed, the new management team went back to basics and PlaNET and its solutions were shelved.

Fast forward to 2003, and Ronnie had come to the end of his contract with Datacraft. Relishing the ability to start afresh, he decided to carve a new path and set up an investment and consulting firm called Interlaken, which both invested in early-stage companies and provided consultancy services to overseas companies looking to set up operations in Asia. Interlaken assisted foreign technology companies keen to explore Asia as a market by providing market research, product localization, business development and strategy. This significantly reduced the risk for the foreign firms and provided them with a better picture of the market before committing valuable resources. Ronnie, and his staff, would then act as the de facto management team to source and close deals. Once the deal was done, companies would have more confidence to come in to their preferred country and set up shop. One such company was Mailkey Solutions, a UK-based email anti-spam and security software firm that was very well-funded. Mailkey was looking for a CEO in Asia that could bring them into various accounts. Ronnie was appointed CEO, Asia, and he helped bring Mailkey into Singapore, Malaysia, Thailand, Taiwan, and Indonesia. A success story, Mailkey listed on the NASDAQ secondary board one year later.

Ronnie also invested in a Thailand-based technology company called MSG Solutions. MSG Solutions had a number of local banks, telcos, manufacturing companies, multinationals, and government agencies as its clients. At the time, they were working on an extremely large contract with one of the largest telcos in Thailand and with the public transport operator. The goal was to provide live in-bus and in-train mobile advertising for the whole of Bangkok using innovative content and wireless technology. He was invited by a few ex-Datacraft employees to invest as they felt the Thai market was huge and underserved. Ronnie became an investor in the company and took over as CEO, developing a vision to build and deploy the next generation of wireless technology solutions using WIMAX (worldwide interoperability for microwave access) technology. He also brought in a number of local Thai investors who assisted in running the local operations and building the business network. Unfortunately, the owner of the telco was ousted in a coup, and the deal was scuttled. As a result of the loss of this large deal, the board decided to voluntarily wind down the company.

As a serial entrepreneur with a wide variety of experiences, Ronnie was the sort of role model the Singapore government wanted to attract to cultivate the next generation of idea-makers. In 2006, Ronnie was approached by a couple of friends, Eugene Wong and Lo Yew Seng, to work with the Economic Development Board (EDB) to set up a business angel fund. The intent was for the government to work with experienced entrepreneurs to assist to co-invest, guide, and mentor early-stage companies. Sirius Angel Fund (SAF) was set up and it was under the Business Angel Scheme, where EDB would co-invest equally with SAF up to S$2 million per investment. The intent was for EDB to work closely with seasoned entrepreneurs to develop early-stage companies that were to be the economic drivers for Singapore's growth.

In 2008, while Ronnie was still involved in Sirius, an old friend, Seah Chin Siong, called Ronnie to see if he was interested in assisting to start up a global consulting and solutions company with the IDA. The firm was to be called IDA International and it would be responsible for packaging Singapore's over 20 years of extensive experience in ICT planning, policy-making, technology solutions, implementation, and operations, and providing these services and solutions to countries around the world. Always interested in trailblazing new fields, Ronnie signed on.

For the next two and a half years, Ronnie worked to put in place a strategy, products, solutions, and a team to engage various foreign governments. These governments were keen to engage Singapore to learn and accelerate their own ICT progress and position ICT as a strategic economic driver, as Singapore had done many years before. But before long, the methodic pace of the public service made Ronnie itch for more action. Keen to get back into the start-up scene, Ronnie left in May 2011 to set up IncuVest, a tech accelerator and incubator with business partner, Natasha Foong. IncuVest is a culmination and combination of Ronnie's 20+ years of experience as a consultant, serial entrepreneur, investor, and government officer. Recently appointed as one of NRF's TIS incubators, Ronnie aims to build a value acceleration ecosystem of sorts and has already invested in six companies to date, ranging from e-commerce, online restaurant reservation, digital media, customer retail analytics, game analytics, and mobile companies. A lasting player in Singapore's entrepreneurship scene, he continues to play in the start-up field through his new role, providing promising technology companies with the necessary funding, structure, management support, and access to markets to accelerate their success.

From his days as a 16-year-old university student to his time as a young entrepreneur in an underdeveloped ecosystem, finding his way in uncertain environments has become second nature to Ronnie. Standing by his convictions kept him on paths he is proud of and he utilizes the wisdom he has gained to train Singapore's next generation of entrepreneurs. From experience, he knows that an entrepreneur's character is more important than his or her idea. Hot ideas need support to refine and become molded. In Ronnie's situation, he did not have an official mentor but he believes budding entrepreneurs can benefit from such a relationship:

> As a VC, as an angel investor, and as a former entrepreneur, I believe my experiences are useful in helping to pull through new entrepreneurs. It is important to support their growth as they are really like the life blood of creation.

Ronnie's satisfying journey was anchored by the power to make his own decisions. In addition to choosing to spin out, to pivot from consulting to selling products, and to sell PlaNET to Datacraft, Ronnie also chose not to accept a number of term sheets from investors whom he felt did not understand his business. Understanding the importance for entrepreneurs of being able to maintain true north, he and his firms are a good example of how entrepreneur-turned-investors better understand the true struggles of building a business. Strong synergy between an investor and entrepreneur forms a backbone upon which clear communication and effective feedback drive the budding business to the next level.

6 System Access and Leslie Loh

Growing up, Leslie Loh often awoke to the sound of a revved engine. Every day at 4 am, his parents would leave their *kampong* in Changlun, Malaysia and drive their truck to the nearby villages to sell the day's freshest fish. One day that might be me, he thought. He and his six siblings were exposed to the lives of such small entrepreneurs who hoped for an easier life for their children. Many dreamt of the day their son would be accepted into university, graduate as an engineer, and become elevated into a different social class. But admission into top local universities was fierce and, as an average student, Leslie knew his chances were slim. With seven children to provide for, his parents could scarcely afford to send him abroad. When he failed his GCE O-level exams and had to retake the year, he found a job as an apprentice in an electrical shop and learned to repair TVs and radios. Accepting that he may never attend university, he hoped for the day when he could open a small shop of his own. Soon afterwards, his father sat him down and insisted he pursue further studies abroad. While his parents had saved enough for him, it was not without financial strain, and Leslie promised to ease their burden by one day providing the same opportunity for his younger siblings.

At that time, many Malaysian students who studied abroad chose to attend British universities because of the similar commonwealth education system. For Leslie, the problem was that such schools had less flexibility in the type of subjects he could select. Where he excelled was in math because he enjoyed it; he was quite weak on all other subjects that he was less passionate about. He liked math and put in the effort to succeed, always ranking near the top of his class in this area. One should only do things that one enjoys, he reasoned, but this conviction would not help him convince Malaysian or British admissions offices to grant him entry, given that they expected grades from a diversity of subjects. Because Canadian schools allowed students more flexibility in course selection, he secured a place in a private college in Toronto to attend Grade 13, the equivalent of GCE A-levels, in preparation to apply for university studies.

In January 1978, the time had come to send Leslie off on his scholarly adventure. To celebrate, the Loh family organized a send-off in Malaysia's cultural hub of Penang. Surrounded by family and enveloped in the warm tropical breeze, Leslie relished the atmosphere knowing how different his life was about to become. Landing at Toronto Pearson International Airport many hours later, his body felt sub-zero degree temperatures for the first time. That year changed Leslie's life.

On the one hand, he was exposed to different cultural norms. He experienced snowfalls, tasted salad and cheese for the first time, and gained the confidence to master a new language. In his hometown, most people rarely spoke English as it was seen as a form of showing off. He would speak Chinese with his Chinese friends, Malay with his Malay friends, and only a bit of English with his Indian friends. Going to Canada, he had only the confidence to say basic phrases like 'yes' and 'no' but the cross-cultural experience opened his mind to different approaches while honing his innate sense of self. On the other hand, Leslie also grew academically. The Canadian education system allowed Leslie to choose his courses and focus on subjects he enjoyed. Besides English, which was compulsory, Leslie naturally decided to choose four math courses and physics as it involved more math than the other sciences. He performed well enough to be admitted into a four-year bachelor of business administration program at St Mary's University in Halifax, Nova Scotia. Halifax's laidback and monocultural vibe certainly contrasted with his home country's tricultural environment! Throughout his studies, however, he kept his eye on the prize of being able to provide his younger siblings with a similar education experience. He chose to take most of his electives in IT-related subjects because he not only grew to enjoy software programming courses, but he foresaw it would be a useful skill to have when searching for a job post-graduation.

Four years later, in 1982, Leslie was armed with a freshly stamped diploma and on the hunt for a job as a software developer. He had initially wanted to start his software career journey in Singapore since it was known as a rising technology hub. As normal in his early years, he experienced rejections even after multiple interviews; perhaps he had not learned to communicate his strengths effectively. Undeterred, he then found a job as a PC software support person with the city's first Apple PC retailer in Alor Setar, the state capital of his home province. Although this was a customer support position, it was exciting work because he was the first to play with the newfangled technology. In the early days of computing, one might need to crawl inside the hardware box to tinker with various dip switches and Leslie enjoyed the hands-on nature of the job. After four short months, he was able to get back on track to starting his software career in the city-state when he was recruited by Singapore Technical Services (STS) to perform a similar position at four times the salary. STS were then the regional distributor of Osborne Computer, the world's first portable PC. Unfortunately, within seven months, the US-based Osborne became a casualty of the PC wars of the early 1980s and went bankrupt, kicking Leslie out of a job.

Fortunately, some of Leslie's customers were impressed with him and job offers came rolling in. But, Leslie thought that none of these jobs would provide him with the platform to write software for the real business world. He did not want to be the super geek in software development, but he felt it would be important to have real-life exposure in developing software in his pursuit of a career in IT. From his time in STS, it was clear to him that there was market need for software solutions for small businesses. But many of these potential customers had to be turned away as the standard software packages, which were limited to standard accounting and word processing, were not able to meet their unique business needs. Traditionally, only the larger companies who could afford to acquire costly mini and mainframe computers would have the financial ability to develop customized software for their specific needs. The arrival of the PC would now for the first time make it affordable for small businesses to acquire customized software that could meet their unique business needs. Leslie believed that it represented an opportunity of a lifetime to ride on that wave. He decided to take a risk and start his own business, focusing on developing customized software for small businesses using the new PC technology.

What did he have to lose? The venture felt risk-free because it would allow him to pursue his passion of writing software, and also provide him the opportunity to acquire diverse skills and gain exposure to the world of entrepreneurship. And if he failed, the experience would give him the software development skills that he could then use to obtain another MNC position.

The next step was to set up his company. Leslie's grandmother loaned him $5,000 ringgit,[1] which was enough to incorporate the company, lease a PC and a printer, and have enough money left for another two months. In August 1983, System Access was officially incorporated. The first step was to find a project that would enable him to survive for a couple of more months. He began by approaching some of his former STS customers looking for opportunities. The owner of a timber-trading company wanted to computerize his invoicing operations. Leslie estimated that such a job would take three to four months to complete and quoted a price of S$1,200. Recognizing Leslie's tenacity, the owner had faith he would deliver. He agreed to the price, and even provided him with a PC and a desk that he could use as a temporary office.

Four months later, the turnkey software Leslie designed was successfully put into full production. It was an incredible feeling of building something from nothing and, in that moment, he felt he could truly consider himself a software programmer as he was able to write something that could be put into practical business use. After completing his first project, Leslie had to move out of his temporary office where he rented an office cubicle from his accountant who had incorporated his company. Leslie was so cash-strapped that he had to pay for his office rental by using his company shares. He jokes that his accountant landlord was probably one of the most successful angel investors in Singapore with the highest return on his investment (estimated to be 180x ROI).

As a one-man software team, Leslie was faced with a steep learning curve putting the limited software skills he learned at school into practice. He had personally to execute the entire turnkey software development lifecycle from functional requirements analysis, to establishing design specifications, to program coding and testing, and implementation of the customized software. He learned to be a multi-tasker simultaneously working on projects while searching for his next customer. Otherwise, the danger was that he would have no income and have to return to the traditional workforce. Despite the hectic and demanding schedule of working 16-hour days, Leslie found it wholly satisfying to be able to enhance his programming skills and learned how to build a business.

System Access' second customer was Tay Liam Wee of Sincere Watch, now a pan-Asian powerhouse retailer in luxury timepieces. At that time, Sincere Watch was a small business and was keen to grow operations. Just back from studying in Canada, Tay saw the need for an integrated software system that would allow his business to manage inventory, monitor sales, and analyze business performance so he could scale his business with standard processes and management control. As typical of entrepreneurs with a small budget, he drove a hard bargain. Leslie and Tay agreed on a price of S$2,700, a true bargain as the software was used for the next five years. Leslie laments that he should have been paid in Sincere Watch shares for his efforts and that would have returned many times over!

With a second customer under his belt, System Access was ready to hire its first employee. Leslie hired a programmer, so he could focus on understanding the customer business, designing the software, and acquiring new customers. Monthly expenditures at this time were around S$3,000, split into Leslie's salary of S$1,200, his programmer's salary of $1,000, and $500 for the rent of his office cubicle. Given that most projects earned around S$3,000 and took three months to complete, System Access needed to have at least three projects ongoing at any one time to keep the company afloat. Wanting more income stability, he took up a part-time lecturing position with the extramural study department at the National University of Singapore. His job was to conduct a software programming course for working professionals titled 'Learning Programming using dBase II'. For a two-day course, he was paid S$1,600, more than his monthly salary with System Access. The position not only provided a consistent income, but also provided an avenue for Leslie to share his passion with a steady flow of potential customers.

The next year, System Access secured their biggest contract yet. It was worth US$20,000 for completion over four months with AQUA Water of Indonesia, now a part of Danone, which supplied filtered water to offices and homes. AQUA required a fairly simple software system that could process collected delivery orders and produce month-end invoices for their customers. While this was straightforward, the PC used was stretched to the limit processing a high volume of month-end delivery orders. Although it took over 24 hours, the customer was extremely satisfied as they had attempted to use their minicomputer for the same tasks but the machine had failed. For this project, Leslie traveled to Jakarta twice a month to understand their business requirements and would develop the software in Singapore for delivery and testing on the subsequent trip. They paid him in cash, so traveling back to Singapore required him to stuff his pockets with US$5,000 of cash. Leslie thought that that he might land in jail 'smuggling' all that cash out of the country.[2]

In 1985, two years and many satisfied and cash-paying customers later, Leslie thought he could not just keep going in the same manner. Although he was not short of customers, he remained cash strapped. The projects did not get easier as each new turnkey software project required a new learning curve with regards to each customer's business operations. Each time, the software would need to be developed from scratch and, once implemented, needed maintenance support that only the original developer could service given limited documentation. Such maintenance support efforts multiplied with the increased number of customers. It became normal to have four to five 'return call' messages every day when he got into the office. Unfortunately, maintenance fees are only charged at 10–15 per cent of the initial licence fees per year, which meant that unless Leslie found a way to charge more or have less maintenance, System Access could not achieve sufficient profitability to hire more people.

With heavy competition given the low entry barrier for the business, Leslie knew he could not charge more as he had many competitors that would be keen to take up his projects. This dilemma had faced many of his peers who either stayed small players or disappeared. Leslie had bigger ambitions for his company. The only logical thing to do was to move from turnkey software development to standard software products, which allowed him to scale his business in term of profitably and value.

The easiest decision was to develop standard products – but for what applications and industry? In the mid-1980s, banks had little choice when it came to software solutions. Banking applications then were developed for large institutions and used proprietary and expensive technology running on mainframes and mid-range computers. Such software packages were too complicated and costly for small offshore operations. Because offshore banks only processed low-volume, high-value transactions, Leslie knew he could write software using the emerging PC platform at a low cost and sought to create the first PC-based banking applications for the large number of foreign banks in Singapore. But without a proven track record as a banking software developer, it was difficult to sign on customers.

With expansive glass windows and stately furniture, bank offices can be intimidating. Luckily for Leslie, the environment was did not intimidate him into cowing down from a daring pitch. 'Let me get this straight', he recalls the managing director of a new offshore bank saying incredulously, 'I have this large international banking software vendor offering me their solution for US$1 million with an implementation time of 18 months. You are saying you can do the same job for S$20,000 at half the time?' As System Access was a two-man company with little experience, he was naturally suspicious. Leslie never gave up and felt that this could be a quantum leap for the company if he managed to secure this prestigious bank as his first banking customer. After several meetings, Leslie convinced the bank with his persistency to take a chance on him – but a chance is all it took. Leslie recalls 'I had promised the MD that I would not take on any other new projects until I got the software successfully installed and working in the bank'.

Working hard to prove his worth, Leslie was the first person in the bank at 6.30 am and the last person to leave at around midnight. After nine months of hard work, he delivered his baby; System Access' first PC-based banking software product was launched. Orders from four other foreign banks quickly followed. This success propelled System Access onto the next level. The company now had a proven track record of selling the software package to offshore banks at S$30–40,000. Since work for other industry sectors only yielded S$3,000 per project, the ever pragmatic Leslie began to specialize exclusively in developing packages for banks. By 1986, the once small start-up was achieving yearly revenue of S$200,000 and had carved a niche in developing software solutions for small, foreign offshore banks in Singapore.

By 1987, System Access had a steady roster of clients. It was a good achievement, but Leslie was not satisfied. The company's current path would ensure their survival but would not make them rich. 'If our competitors can sell their solutions for seven figures, why can't we?' he reasoned. Scanning the industry, he noticed there was a move towards the adoption of open systems technology. In the mid-1980s, a new computing paradigm – combining UNIX operating system, TCP/IP protocol, relational database, and client server architecture – delivered a more cost-effective alternative. For the first time, System Access could develop their software on cheap single-user PCs and deploy it on larger multi-user mid-range computers.

The catch was that redeveloping the single user software into a multi-user system would require a significant learning curve and investment that would potentially take two to three years. To some, it appeared to be a step backwards – redeveloping technology versus gaining new clients – but Leslie foresaw that the bet would be worth it. If he was right, the company could potentially leapfrog international competitors by delivering one of the industry's first banking software solutions able to run on an open systems technology platform.

Armed with a brief proposal, Leslie approached the National Computer Board (NCB) for partial financial assistance to develop a new generation of open system-based banking software products over the next two years. There, he met Saw Ken Wye, now a senior executive in Microsoft Asia, but then the young officer who evaluated his proposal. He explained that System Access could potentially change the banking technology landscape with five banks ready to adopt the product once ready. 'If NCB is willing to provide partial funding support, it will give me the credibility to convince these banks to fund the rest'. After receiving in-principle approval from NCB, Leslie approached his current customers and convinced them to invest as well. He pitched:

> Your business is growing and you will need a new software solution that will support more than what our current PC-based system could support in the near future. Rather than going out and paying more than S$1 million for an aging proprietary solution, we will develop something similar but on a new generation of technology at a fraction of the cost. You will only need to pay 10 per cent of our total development cost. We have strong support from NCB on this plan.

The result was the first version of the award-winning SYMBOLS software launched in 1989. Ultimately, this became System Access' flagship product and was the foundation that propelled the company into the future. Leslie believed

SYMBOLS was possibly the first treasury banking software product in the industry that adopted the then advanced client server and relational database technology. As a small player, a superior product was fundamental to the company's success. Since 1987, the company had steadfastly invested more than 1,000 man years of product development to evolve SYMBOLS from a departmental treasury banking software to one of the most technological advanced and comprehensive universal banking software products in the world. As a result, the pricing of the product jumped significantly from S$30,000 per customer in 1990 to more than US$100 million from the largest customer in 2005.

How did the company make such a dramatic leap? Looking back on his journey, Leslie pinpoints specific opportunities that allowed System Access to make quantum leaps. That is, he took risks that successfully propelled the company to more advanced levels. While embroiled in day-to-day operations in 1989, Leslie received a call out of nowhere. The caller represented one of South Africa's largest banks and had heard from a connection in Italy that System Access had the software they wanted. 'That's strange', thought Leslie as he had no Italian customers. What happened was that System Access had approached an Italian firm's Malaysian office and word had spread over the grapevine. The caller pressed, on explaining he had searched the globe and only two firms – System Access and an Irish firm – had most of what they required. Despite their small size, they were the only ones with the technology. He invited Leslie to South Africa to talk about potentially customizing the software for them.

The problem was that, at the time, Malaysian nationals were not allowed to enter South Africa and as a small business the price of a flight ticket looked steep. Instead, Leslie explained that he was very busy but would be happy to welcome them if they wanted to come to Singapore. They made the trip and Leslie not only demonstrated the capability of his firm's technology but was a prime example of Asian hospitality driving them to visit Malaysia's cultural hotspot Melaka and introducing them to Nyonya cuisine. By the end of the week, they had become genuine friends and promised to keep in touch. Shortly afterwards, they presented their review of System Access to their management board who were greatly interested in meeting Leslie and having him demonstrate usage to the staff. They asked him to come for three weeks, and both bought him a plane ticket and worked things out with the customs department. With a friendly nature and disarming chuckle, Leslie and his demonstrations were very popular.

Towards the end of his trip, he had lunch with the managing director who asked him his price. Leslie took a risk and, despite selling his past projects at a maximum of S$150,000, quoted a price of US$2 million for 18 months. The risk was that the MD would have been offended by his audacity – but he simply asked for justification. The quality of the technology spoke for itself and obtaining a project of this nature would catapult System Access into a different realm. Honestly speaking, he explained that that the value of the technology was much higher for them than for a small firm. Together, they worked out that the current technology only met 70 per cent of their needs. The bank would hire six Israeli engineers out of their pocket to customize the software. Agreeing to Leslie's price, they had but two conditions. The first was that they wanted the source code at a time where all players kept this closely guarded. With System Access being a small company, protecting their investment was of high importance should they become bankrupt. The second was that they wanted Leslie to be the project manager, despite his position as managing director of System Access. Leslie had reservations about both, but he says it did not take him longer than a few minutes to agree to their demands, excited as he was that System Access had made a quantum leap obtaining their first large international customer. Should they be successful, they would become a different company – so he gambled his entire future on the project. With a good track record, he knew the company was on their way to attracting VC funding and future customers.

With a taste of the global market, the company focused on how to expand their reach. Simply put, established markets would not take a chance on emerging market technology no matter how sophisticated it was. Singapore's domestic market was also too small a pot for the company to achieve scalable growth. System Access had no choice but to globalize. 'We needed to enter markets where our advantages would be best appreciated and where we could compete favorably', says Leslie of their strategy. 'We needed to choose a battlefield where we were best able to compete. The largest market means nothing if you cannot win'. To support this plan, the company accepted S$2 million in venture funding from Nomura Jaffco and EDB Ventures in 1991.

The first step was figuring out their niche differentiation point. Their banking products were based on a new generation of technology – integrated solutions that could support the end-to-end business operations of banks. Clients that would most appreciate their products were those with the minimum legacy infrastructure constraints that could support the adoption of the technology. Initially, they targeted banks in emerging markets including Eastern Europe, the Middle East, Africa, and South Asia. In these regions, large international players and small regional players were on the same playing field, since there banking software solutions were not yet available in the market. Leslie sought to leverage on Singapore's favorable reputation in emerging markets as a key financial hub. Gradually, System Access became known as a company with advanced banking knowledge that enabled them to deliver banking software products capable of supporting international banking practices.

By 1993, System Access had gained enough momentum to set up their first international office in London, albeit on a shoestring budget. Barry Gilbert, a British national hired by the Singapore office, was transferred there to expand the company's presence into eastern Europe. Two years later, Gilbert opened an office in Dubai to extend market coverage to include the Middle East and Africa. Their go-to-market strategy was to leverage on the global distribution channels of their technology partners, including Oracle and Unix-based hardware vendors, to target customers in a cost-effective and efficient manner. As producers of the only Oracle RDBMS[3] Unix-based banking software product at the time, System Access was highly sought after by large multinational technology vendors. These vendors saw the technology as able to complement their Unix-based hardware and

Oracle databases in delivering integrated technology to their network of banking customer. By the end of 1998, System Access had annual revenue of S$12 million and penetrated 15 emerging markets attracting the attention of a top US VC firm.

Just before the height of the dotcom boom in 1998, System Access received a significant investment from the large US venture fund, Warburg Pincus. They saw that System Access had emerging market reach and superior technology; with the rise of internet technology, they had an opportunity to become a global leader. Leslie believed they needed US$10–15 million to grow the business and were prepared to give up 30 to 40 per cent of the company. Warburg Pincus counter offered US$25 million in exchange for 50 per cent of the company. Mulling it over, Leslie agreed but asked for US$5 million for his personal use and US$20 million to grow the company. This way, should anything happen to the company; he would feel his efforts were worth it. They agreed. At the time, this was likely the largest amount obtained by a homegrown software firm and was significant because US investors tended not to invest overseas. With high stakes in the company, the VCs naturally became heavily involved in working with Leslie and his management team.

During this period, companies were valued almost entirely on their growth potential with limited references to their actual financial performance. Leslie vividly recalls one conversation with his investment bankers on how he could rapidly increase System Access' valuation. They said:

> Leslie, your company could be valued at US$1 billion in a couple of years if you do the right things. You need to invest significantly to grow your revenue by capturing market shares, particularly in the United States, where your core investors will reside. It is fine if it means that you will not be profitable in the short term. You need to recruit and showcase an A-team comprised of experienced management professionals from large organizations who have done this before.

Given the euphoria of the dotcom era at that time, the advice seemed to make sense and Leslie took the advice to heart. He set out to enter new markets and build an A-team that would be acceptable to the US investor community. To better position the company for a public listing on NASDAQ, System Access extended their current focus on emerging markets to target developed economies including the USA and western Europe. He lured a number of MNC stars from top companies by tempting them with huge stock options and the guarantee that they would all become multimillionaires if System Access became publicly listed. Revenue growth was targeted to increase 50 per cent each year, but System Access was willing to compromise on profitability in exchange for a significant investment to drive sales.

Danger signs were starting to appear, but they were masked by anticipation of stellar growth. Though System Access expected to have a US$5 million deficit for the year, they remained on track for a public listing during the summer of 2000. A leading global investment bank had been appointed as its underwriter and anticipated the company would be worth US$800 million. Leslie was sitting at one of Singapore's landmark hotels when he received the valuation from his investment banker. Surrounded by marble floors and luxurious flowers, he stammered 'Are you sure this is for real? You mean my little company is worth more than this luxury hotel that has been around for ages'?

In April 2000, the company hit a huge speed bump in the form of the dotcom crash. Overnight, technology companies lost their reputation as the darlings of the stock market. The burst of the dotcom bubble brought the industry back to its senses and companies in the red were no longer welcomed by investors-at-large. In a rush to grow the business, System Access had built up a high cost structure where sizable investments had been committed in the short term for operations that would only yield financial returns in the long term. A complete overhaul of the cost structure was required with a complete revamp from pursuing top-line revenue growth to a bottom-line growth strategy. The company needed to retreat from the developed economy markets where they were not competitive and unable to deliver quick financial returns. The key challenge lay with how the company could cut business costs while still delivering sufficient cash flow and continuing to invest in building capabilities for future growth.

It was shocking to have reality sink in after being catapulted to the big leagues. Over the past couple of years, he had been playing to the gallery in his attempts to meet the volatile expectations of the investor community. With too many hands in the pot, he lost his direction and began to wonder if he had the right experience and leadership to take on the overwhelming challenge of turning the business around. After some deep soul-searching and discussions with his venture investor, Leslie decided in 2001 that it would be best to bring in a professional CEO to lead the company's day-to-day business operations. He and the board agreed to hire a CEO who had experience leading a multi-billion dollar business unit for a global technology vendor.

Although he was no longer CEO, Leslie remained as chairman of the board and involved with strategic guidance and corporate governance. However, execution was left to the CEO and his professional managers. Although the change took some getting used to, Leslie believes that taking a step back gave him the opportunity to observe operations and it was one of the best learning experiences of his life. He credits the experience as changing him from a start-up entrepreneur to a capable entrepreneurial leader saying:

> For the first time, I was not the captain on the field initiating and being involved in almost every move. I was an observer watching from the sidelines with a bird's eye view, witnessing the implications and effectiveness of every move executed by the team.

For System Access' US investor, their money had by 2003 been simmering for over five years and they wanted to explore an exit. An investment bank was appointed to source a buyer and, within months, a leading competitor made an offer of US$23 million. Had he agreed to the sale, Leslie could have walked away with a very comfortable sum but he decided against it. What he wanted to do was fix the issue and find an exit he felt the company deserved. 'Give me a bit of time and I will raise the money to match the offer on the table', he told the investor. Understanding Leslie's passion for the company, he gave him six weeks.

The following weeks were an intense whirlwind with Leslie meeting over 100 potential investors. Bleeding with determination and even investing US$2 million of his own savings, he convinced a consortium of 40 investors that he could turn around the company's deficit in one year and arrange for a public listing by 2005. Together, Leslie and his staff owned 40 per cent of the company with the new group of investors owning the rest.

Having secured a controlling stake in the company, Leslie resumed his position as CEO and worked to turn the business around. The A-team had inculcated a culture with structured processes led by professional managers, but the old culture that was lean, mean, and hungry for success was gone. Leslie was determined to motivate his team and get that feeling back saying:

> When we were a start-up, I was the captain on the field but also its forward and defender. I was involved with every move and the company could not progress without me. In the transitional stage, I needed to be more of a 'player coach' where I would deliver strategy, mentorship and guidance from the sidelines but would also inject myself into the game as player when required to deliver optimum results.

Teams excel when all members breathe their mission. Skills and expertise are important, but the secret ingredient for success is heart. Hunger and passion cannot be taught, but management skills can. Back at the helm, Leslie needed such support and he replaced those who no longer had excitement for the role by promoting passionate team members with little management experience into senior management roles. With his army behind him, Leslie ventured into battle. One year later, in 2004, he succeeded in turning the deficit into a S$6 million profit and began to prepare the company for a public listing in 2005.

Years of blood, sweat, and tears paid off when System Access reached the big leagues. That July, the company was successfully listed on the Singapore Stock Exchange. Annual revenue was hitting S$30 million but without gaining headway in new markets, growth would plateau. Leslie decided to pivot focus on high-value banks in the developed market. The David in a sea full of Goliaths in the form of Oracle, Tata and Infosys, this decision was made having already beaten the latter in emerging markets. Unfortunately, it was simply too risky for large banks to take a chance on a small vendor with limited resources.

After careful consideration, Leslie realized he had taken the company as far as it could go. It would be impossible to grow organically to the size of their Goliath competitors. Finding the right company to adopt System Access would take the company through the next phase of growth. Six months after they were publicly listed, Leslie kicked off the mergers and acquisitions process without engaging the services of an investment bank. He knew every nuance of the company and knew how to position the company for higher value. System Access also had the network to reach out personally to players that would be better able to leverage their value.

One month later, the company received expressions of interests from seven large MNCs from the USA, Japan, and India. Leslie was not surprised saying proudly, 'Their interest did not come out of the blue. I have always maintained a certain level of friendship with our partners and competitors in anticipation that we would one day pursue inorganic growth together'. Even potential buyers with whom System Access did not have a direct relationship would pick up his call given the company's profile.

In July 2006, SunGard, a US-based global leader in the financial services software sector, acquired System Access for S$120 million. At triple the market value offered six months earlier, Leslie was right to rebuild the company and steer it back to its true north. Given their strength in high-growth emerging markets and the large core banking systems market, SunGard was happy to pay a competitive price. The final bidders were all large competing vendors, creating a competitive bidding landscape.

Looking back on over three decades of entrepreneurial experiences leaves Leslie with a sense of wonder. Hard work and tenacity pushed him to seek out and take a chance on quantum leaps, which he sees as one of his secrets to success. No one could accuse Leslie of being in the right place at the right time. As a fresh graduate, his first employer folded and he found himself jobless in an adopted city. Later, his company was hit by the NASDAQ crash of 2000 just as they were preparing for IPO. A willingness to stick through the tough times and rebuild characterizes Leslie's success. Certainly there were openings where he could have thrown in the towel and cashed out, but he believed in his company and knew he could change the situation. Since childhood, a stubborn dedication to an innate sense of self carried him through morale-bruising schooldays. He was certain that all he needed was to be excellent in a few areas he was passionate about. Looking at his history, no one can deny how well things have worked out as a result of his dedication to stick to his true north. Leslie's story is a great example of the value of believing in young entrepreneurs, especially those with a clear focus or those strong enough to break away from tradition. Leslie's Singaporean landlord from his early System Access days is certainly glad he took a chance on the budding entrepreneur. Allowing Leslie to pay for rent with shares of his young company, he eventually received 180 times his initial investment. 'The landlord with the best return in town', says Leslie, smiling.

7 Systems@Work and Ng Fook Sun

Recently, strategically placed walls at many of Singapore's MRT stations have become one-dimensional shopping malls. Customers clamor for hot deals such as one-for-one *Wicked* tickets from Showbiz Asia or 5 per cent discount of the quintessential satchel from popular shoes and accessories company Charles & Keith, a local brand. The advertisements are placed in areas with heavy foot fall, where shoppers can easily take advantage of deals by taking out their smartphones, scanning the QR code linked to their preferred product, and checking out right then and there. Today, technology is enabling sellers to connect with buyers where they are. As one of the world's most plugged-in cities, Singapore is a hotbed for testing mobile payment technologies. One decade ago, smartphones were not nearly as ubiquitous – but a visionary local entrepreneur was already anticipating the potential of the mobile payments market, more than a decade ago.

The son of corporate professionals, Ng Fook Sun left his home country of Malaysia to attend Singapore's top university, the National University of Singapore. Graduating with a degree in electrical engineering in 1988, he pursued a path common to many high-achieving graduates. While he loved scholarship and originally planned to pursue a PhD and stay in academia, he accepted a development position with the prestigious corporation Singapore Technologies (ST) and began a six-year stint. His positions enabled him to travel around the world, namely to the USA and Japan, satisfying his desire for global exposure. After six years, he decided to join a US-based company affiliated with ST in a business role. However, corporate restructuring soon pushed Fook Sun to seek an opportunity with Gerber Corporation in a regional managing director position. By this time, he had risen to run the APAC office working with precision equipment and software for the industrial sector, but with headquarters in the USA, he constantly reported to a senior vice president and the CEO. The dawning realization that he was responsible for the day-to-day but had no control over major decisions began to gnaw at him. He decided to make a change. When he looked at what he enjoyed most about his past jobs, the emerging field of IT stood out. Sure the industry came with grief and headaches, but it was the field that excited him most. He liked the constant change and began to look for a way to get back in.

Fook Sun did not see leaving his comfortable MNC lifestyle as a risk, but moving to the uncertain world of an IT start-up is atypical of his personality. Entrepreneurs are often perceived as hyper-confident and intense beings, with their passion steaming out of their ears, and Fook Sun's decision belies his under-the-radar personality. He might take some calculated risks but he does not have a penchant for spontaneous leaps. As a humble hardworking local boy, Fook Sun is a good role model for those who do not inherently possess an audacious persona. Surely his journey is a testament that success results from careful planning and execution?

Surprisingly, this is not the case. 'Foolishly not', he said when asked whether he had the idea for his new company before leaving his MNC job, but the decision was matched with a focus on entering a bulletproof industry. In Singapore, there are a few stable industries where companies and consumers alike have big budgets. One is education – the city is flooded with tuition centres and bookstores packed with practice tests. Another is banking with sustainability and growth of paramount importance. It was a close friend and soon to be co-founder that turned him onto the financial industry. Together, and with another two friends, one from childhood, and the other with success in starting up one of Singapore's earliest technology companies, the group conceptualized a new business around electronic payments for the growing e-commerce and mobile commerce markets. The group was certain they could engineer a system that would provide a unique proposition to fulfil the requirements of these markets. Fook Sun had thought about moving to another corporate job to try and pursue similar technology development work but relatively quickly concluded that this was a dead end as there were few development positions at that time in an industry that was very young.

The caveat was that Fook Sun did not have experience in the financial sector. Instead of worrying about opportunity cost, he channelled his efforts to focus on execution. Ever resourceful, his first step was to build a team of experts. A team crystallized over a supper with his soon-to-be co-founders when the conversation flowed to mobile payments. Between the four of them, they had experience in the financial and consulting industries and with the domestic payment network in Singapore but, more than that, after years of working for MNCs, they all shared a vision of wanting to create an environment that was very non-hierarchal and that would make 'everyone motivated to turn up for work every day'. In 1999, their chemistry was made official through the launch of Systems@Work, with Fook Sun as CEO. Their business model provided software as a service with very low-cost licensing. On a pay-per-use or fixed monthly fee, they provided their initial clients – such as EZ-Link, Comfort Taxis, the Land Transport Authority, and the Development Bank of Singapore – with a payment service for processing e-commerce and other online electronic payments. The early 2000s was an exciting time of technological volatility and industry uncertainty and many forward-thinking start-ups suffered at the hands of an underdeveloped market.

At the turn of the century, Systems@Work already saw the potential of applications that glue banks with end merchants. Merchants are able to push more volume, with revenue generated through a pay-per-use model. For example, customers wanting to purchase TV episodes on demand would be able to do so in a secure environment; no longer would they have to wait for their preferred choice to appear on the cable network's rotation schedule. They would subscribe through a company's official website but be redirected to Systems@Work's secure application when the time came to pay. While business became steady, with fees either paid by the merchant or by the consumer as a built-in fee, the company did not gain the kind of momentum that attracts large investments. Many investors did not understand the nature of the business or were simply not interested in taking a chance on the technology.

Like its CEO, Systems@Work stayed focused. For its first eight months of operation, it generated no revenue. How did the company survive? As a technology company committed to creation of intellectual property, the company required significant funding from the onset. Fook Sun remembered thinking, 'Technology companies strive to create intellectual property. In order to do so, we needed time – runway to develop and create value'. Thankfully, this was solved with the participation of a very committed and visionary co-founder and financial backer. Fook Sun offers these words of wisdom:

> If you want to build IT and you want to develop products, there is no way you can do it without rocket fuel [drive and funding] to push you through the development phase. If you can do it in two months, so can the next fellow and the IP will be very shallow. If you are going to build a system, you need to invest some time in it. That is one of the things I find rare in this industry; a lot of people have no patience for that.

With a focus on quality over quantity, the company gained momentum and began to seek funding to power them through the next stage of growth. At that time, there were a number of progressive government grants. Obtaining some of these helped the company gain momentum, but came with their own set of shortcomings. The reality of start-up life is that employees often wear multiple hats. One employee could be taking on responsibilities that would require three MNC employees to execute. However, the grants only allowed the company to claim for as many people as they employed, which, while it made sense technically, did not reflect the true cost of development or the nature of the work being done.

To mend this funding gap, the company worked hard to generate sales and cash flow from the business. It did not accept any other investment other than from its original investors. Why did they not form a partnership with VCs? Fook Sun admits that he 'tried like hell'. The company actually started out on the traditional 'Silicon Valley' notion of funding going through seed, Series A, and then Series B funding. Originally, he was very positive believing that if there were enough VCs in the market to create market parity, it would be rare to receive a sour deal. After all, a rising market translates into a good deal for both the investor and entrepreneur. After gaining momentum with previous funding from the co-founder/financial backer, the next round should have been attainable. After two years of fundraising but with no success, Fook Sun became so disillusioned by the situation that it fired up his survival instinct and motivated him to succeed without ever having to rely on others again. The problem lay with an underdeveloped ecosystem. Systems@Work needed funding to build the company, but in the investors' minds there is no money behind deep R&D. At the time, VC firms were mostly late-stage investors and pursued general overspecialized industries because of the small market. They compared the risk and return of investing with financing for listed companies – although the rate of return may not be as high for the latter, the floor risk was more comfortable.

Then in 2005, the company's market window shut down, further challenging the company. By then, it had become clear that no further significant funding would be available and that the only way forward would be to focus the company on generating revenue, to quickly reduce the cash burn-rate and turn the company cash-flow positive. Changing direction to focus on the gateway business seemed the best way to utilize their strengths and take advantage of what they believed was a promising growth area. They were not alone in their foresight and, over the next five years, the market became very crowded. Competing as a small player with just a few dozen employees was a challenge for Fook Sun but he emerged victorious because of one simple principle: perseverance. This did not come naturally; engineers hate uncertainty but he was adamant that Systems@Work bulldoze through the tough times and not give into the temptation to give up. Although he was the underdog, he was motivated by the desire to build a quality company irrespective of quick fame and fortune. His grit paid off when the market consolidated a short time later. Many of the former market leaders fell off the radar, failing to innovate and adapt with the market.

Small players have the advantage over MNCs in adapting with an emerging market because they thrive on employees' ability to think outside their scope. Flexible processes then allow easy collaboration and experimentation. Despite fresh momentum, Systems@Work ran into obstacles scaling the business because of banking laws. Southeast Asia may have a population of 600 million but the banking industry is highly fragmented. For example, Singapore banks can only process Singapore-registered merchants, Malaysian banks can only process Malaysia-registered merchants and so on. Fook Sun continued to explore other ways of growing the busines, but, by this point, it had become a labor of love. When I interviewed him in 2009, Fook Sun confessed he was content with steady growth and would be happy working on the business for the rest of his working days: 'To be on the receiving end of an acquisition changes the parameters, and it may seem to be attractive financially, but I am not sure whether that is the path I want to go down'. Two years later, the sky dropped him a pastry ('天上掉馅饼').

One of their clients was a listed German financial services and technology company called Wirecard AG. Essentially, they were in the same business, running payment gateways and developing applications in-house. The difference is that compared with Systems@Work's revenue of S$5 million per year, Wirecard earns S$500 million per year; having the market size advantage of the

Single Euro Payments Area, the Eurozone banking system that links 17 countries through one compliance law. In 2006, they had even bought a bank to vertically integrate the market chain into their business. With great success in Europe, they had in recent years, set their sights on Asia. Singapore, with its strategic location and developed infrastructure, was designated as a key destination and they wished to accelerate traction. Systems@Work had what they needed in terms of clientele and market reach in the region. Fook Sun and his co-founders were not even thinking of selling the company when they received the offer in 2011. The initial surprise gave way to excitement after remembering a conversation Fook Sun had with old friend System Access's Leslie Loh. Over drinks, he had once told Leslie that he was never going to retire in favor of growing his company for as long as possible. His friend encouraged him not to be so stubborn and to create demand for his company. From experience, Leslie knew a company is worth as much as their acquirer is willing to pay and that when such an opportunity presents itself, it should be seriously considered.

A visit to Wirecard's headquarters convinced Fook Sun that the company would be the right home for Systems@Work. Located in a key economic region, Munich has a history of attracting ambitious and hardworking people that are drawn by high quality housing, excellent public transportation, and a low crime rate – much like Singapore. Impressed by the company's watertight processes, he and his team agreed to start due diligence once they returned. Similarly keen to further the process, Wirecard sent their legal team to Singapore for a three-day visit shortly afterwards. After just one day, they finished their assessment, as Systems@Work's books were simple. They had no inventory or loans, areas where there is traditionally worry about write-offs. Speeding along, Wirecard decided to skip the term sheet and go straight to the contract. Here, together with his co-founder, Fook Sun presented their expected settlement amount. They used no scientific formula to come up with the number, but it was an amount the founders could accept with no regrets. Wirecard then worked with them and remarkably the whole acquisition process completed in an exceptionally short period, with both sides focused on the common outcome of consummating the deal. Business does not stop for preplanned vacations, so Fook Sun was sent the contract the very day he walked off the plane for a ten-day family vacation in the USA. Through ten days of exploring the twists and turns of San Francisco, he was constantly in touch with his lawyers. At 8 pm the day he returned home, together with the co-founders of Systems@Work, he signed the contract. Clear lines of communication ensured both sides came out of the whirlwind process satisfied. 'As a technology company', he muses, 'you always wake up unsure of what is going to happen to you'. For Fook Sun, he was proud that his humbling years of sticking to his true north had reached a satisfying conclusion.

Looking back on his 12-year journey, Fook Sun has a few nuggets of wisdom for today's aspiring entrepreneurs. Ever humble, he insists that he is not a successful entrepreneur, but is content in being open to new experimentations. Perhaps this is a hallmark of a truly successful entrepreneur. Perseverance and an openness to constantly learn helps entrepreneurs soak up fresh perspectives. A great amount of success in any event is timing or luck, but such a mentality can increase the probability of this happening. When an entrepreneur steps out of his or her comfort zone – especially when, like Fook Sun, he or she pursues an unfamiliar domain – keeping an open mind reduces suffering. That the majority of the founders of Systems@Work were not in the payments industry from the outset was a big setback when they were starting out. He admits: 'If you asked me to enter the payments industry again today, knowing nothing as when I started out, I wouldn't do it. In retrospect, I think the risk we took going into the market was crazy'!

The company only truly found its footing after seven years of tinkering and learning at a high cost. Reaching their goals in terms of the ideal customers, industry, and position took two to three times longer than expected. He was dismayed by the amount of time it took him and his team to grasp the basics of their adopted industry but decided to persevere and it paid off. Taking the time to talk to people in the industry and hear their uncensored opinions proved invaluable. Sieving through all the various comments, they were able to mine enough gems that proved significant in chartering the course for the company.

While it may appear to be an oxymoron, Fook Sun proves it is possible to be a cautious risk-taker. Early frustrations raising capital honed his survival instinct and his perseverance resulted in success. Having exited from his first and only start-up experience, he remains in Singapore's entrepreneurship scene as an angel investor with Red Dot Ventures, helping a new generation achieve their dreams. Those wanting to enter the mobile payments industry can rest assured Fook Sun still believes mobile payments are the future. The ubiquity of smartphones has formed a solid platform upon which mobile payment applications can flourish. The main issue now is lack of a traction-gaining application. Most important is the speed factor – the app must enable a payment transaction that is faster than cash with today's productivity-minded generation favoring an efficient transaction experience. Of course, most retailers want to shorten queues and improve the experience of shopping at their outlets. Perhaps retail benefits – such as incorporating loyalty cards, discounts, and promotions – or another consumer deal-focused strategy will drive mobile payments. It remains to be seen whether the industry will produce a visionary able to create a disruptive standard for mobile payments like PayPal has done for e-commerce. Who will be this generation's Fook Sun?

Lessons learned and key takeaways

Build a sustainable and valuable business by funding the venture with customer revenue, and delay venture investments until you have a validated and scalable business model

This is the most important theme in Part II. All four technopreneurs – Chak, Ronnie, Leslie, and Fook Sun – did not waste time and effort to raise venture capital at the start for building the technology portion of the venture and plunged straight into the market with their know-how. Each built a viable, sustainable business through a solutions approach, initially from use of off-the-shelf hardware and software and built the integration/customized applications to satisfy customer requirements. Within a few short years, the nature of technology business pulled all four, in their respective ways, to change to an innovation or product-centric strategy to continue their differentiation, and it is through this important pivot to a scalable business model that enabled them to become attractive acquisition targets for larger companies. Chak, through his extensive networks and close partnership relationships with technology companies, was able to strengthen his solutions-only company by merging with other like-minded technopreneurs' more product-centric companies to form Finesse Alliance, which eventually was acquired. Ronnie got tired of his personal involvement in each systems integration project and was able to pivot PlaNET into a pure-play product company with AIMS and received a lucrative offer from a larger corporate entity. Leslie spotted the market opportunity for SYMBOLS, System Access's flagship software banking product, after he analyzed the market gaps for financial applications and built a US$100 million company from this opportunity. Chak managed his venture investors deftly and was able to engineer a win-win exit for all his stakeholders. Leslie raised VC funding only after he has achieved a certain scale and was able to receive, what was then, the single largest VC investment in a local software company. Ronnie took the opportunity of the dotcom boom and parlayed his venture into an attractive acquisition target. Fook Sun never wasted his efforts into raising venture capital. He focused all of his efforts into building a viable business via customer revenue and was able to sell his company after twelve years. One key insight emerges from this series of stories and from the narratives from Part I – taking venture capital has the net effect of accelerating not only your firm's success but also its failure, as contrasted to the longer period afforded to organic growth entrepreneurs to create the exits on their own terms.

The technopreneurs all had the same dilemma:

> raise venture company in the early develop stage of the company to pursue a innovation and product-centric strategy but dilute your control over the company's direction before you have reached traction or validation.

Or:

> maintain control over your company's destination by first building a consulting-based business and pivot to a technology company later.

Having external investors with a strong board voice will accelerate your company's progression or regression – they will introduce you to new partners, induce board-level decisions relating to mergers and acquisitions, set high-level strategic directions, intervene in the company's selection of executives, and other important activities that can bring you a quantum leap in performance or bring you closer to the precipitous edge of bankruptcy for failed experiments or implementations. Many of the success stories in Singapore seem to center on the self-built, organic growth entrepreneurship, leaving the Silicon Valley model of entrepreneurship, where the idea of accelerated failures is the key to success, as a theory to be tested in Asia. We shall elaborate Singapore's effort in emulating the Silicon Valley success formula in Part III. With the successful exits of many technopreneurs who started their firms in the mid-1990s, Singapore now has a local crop of experienced technopreneurs eager to be the coach in the new game of entrepreneurship.

Part III

Coaching the young

Modern Singapore's new crop of funding schemes and incubation programs

With sky-high valuations from risk-taking investors, Silicon Valley continues to attract the world's most opportunistic entrepreneurs. Over the past decade, the world has seen tech companies from this region change the basic building blocks of how societies around the world live, think, and communicate. The Silicon Valley ecosystem and the advent of new technology ventures actually emerged in the 1960s. In 1957, a group of engineers known as the 'Treacherous Eight' left Shockley Transistor to create Fairchild Semiconductor Inc. with funding from Silicon Valley's pioneer VC Arthur Rock. Former Fairchild employees subsequently spawned many entrants to the semiconductor field including Advanced Micro Devices, Intel, LSI Logic, and National Semiconductor. These spin-off firms in turn spawned more firms including Cypress, Zilog, and Sierra Semiconductor. The Fairchild model attributes the spin-off phenomena to employee learning and a positive external environment conducive to new venture creation and is often used to describe the high technology cluster development in Silicon Valley and Boston's Route 128 in the USA. Through successive iterations of start-up successes and failures, Silicon Valley has evolved to contain a thriving ecosystem involving many facets of entrepreneurship: entrepreneurs, investors, incubators, universities, technology labs, technology giants that were the beneficiaries (such as Google, Apple, and Oracle), lawyers, accountants, banks, and other professional firms and networking organizations. Silicon Valley remains the epitome of a vibrant high-technology start-up ecosystem and is the model of innovation success for many countries, including Singapore, to try to emulate and replicate. What forms Silicon Valley's success is complex and dynamic, and would be difficult to replicate without understanding its underlying societal demographics. Figure III.1 highlights the vital differences between Silicon Valley and Singapore in entrepreneurial propensity. Although this figure shows Singapore, you can generalize this model for many other Asian countries and show similar patterns. Ironically, Singapore was also very entrepreneurial during the 1960s, concurrent to Silicon Valley's development. The major difference was that back then in Singapore, entrepreneurship came in the form of the trading-style or bricks-and-mortar type of small family-owned businesses to create income-earning opportunities in an underdeveloped economy. In contrast, the motivation for entrepreneurship in Silicon Valley was driven by learning and the desire to do something new or different than what was in the large corporations.

The first technology start-ups from Silicon Valley were formed from spin-outs of large technology firms such as Fairchild (and later Intel) or research labs such as Xerox and its famed Palo Alto Research Center. In the mid-1990s, Singapore had the wealth composition close to that of Silicon Valley, but paradoxically, the wealth had diminished the drive for entrepreneurship – partly due to the societal emphasis on stability and parental pressures for the young to obtain places in more prestigious jobs. Working for the government, multinationals, and in other 'safe' careers were valued highly and the social norm. The truly gifted and the elites (both in terms of education and family resources) were encouraged to become doctors, lawyers, and other high prestige professionals.

To overcome this difference in social development, one key ingredient is social acceptance of entrepreneurship. Indeed, the USA not only accepts entrepreneurship, but regularly celebrates the heroes and innovators from high-tech, with or without government support: this is a critical societal difference.

I have been teaching technology entrepreneurship to Singapore's university students continuously since 2004 through my involvement with NUS's Overseas College program. Anecdotally, the students from the 2004–2009 year groups seemed to favor banking careers as those paid the best. Keep in mind I only taught students who have self-selected themselves to earn a technopreneurship minor with the requisite required coursework and internship in start-ups. Interestingly, from my numerous contacts with young entrepreneurs in Singapore today, it would appear that the parental pressure to join a 'safe' career is diminishing, perhaps with the family's increasing affluence and the longer time span to memories of poverty, combined with recent strong government emphasis and

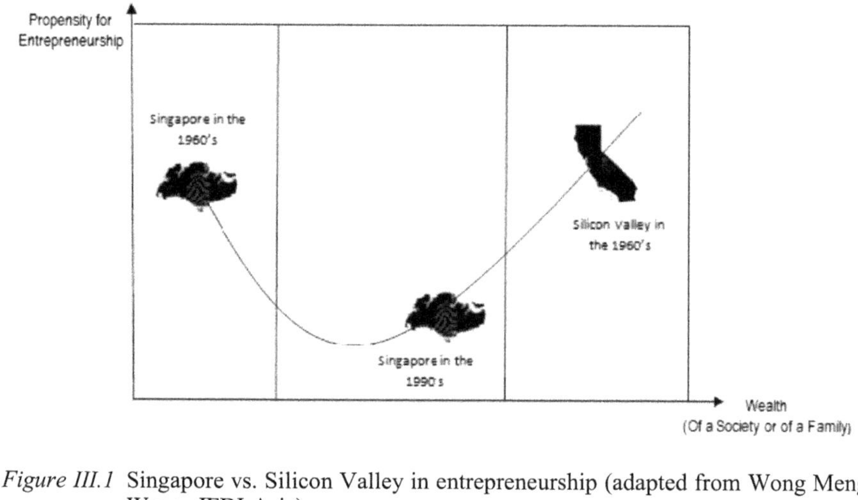

Figure III.1 Singapore vs. Silicon Valley in entrepreneurship (adapted from Wong Meng Weng, JFDI Asia).

support on innovation and entrepreneurship. This anecdotal observation is supported by a 2011 Global University Entrepreneurial Spirit Students' Survey (GUESSS) survey that showed that 37.5 per cent of Singaporean students were interested in pursuing entrepreneurship and have considered starting their own business.[1] In comparison, 36.3 per cent students from OECD[2] countries indicated similar interest. In the same survey, 55 per cent of Singapore students had entrepreneurial aspirations (five years after studies) in the longer term, which is significantly higher than the OECD average of 42.6 per cent. With the proper building of the various facets of the ecosystem, we should see more entrepreneurs joining the technology start-up scene locally and not relocate to Silicon Valley to fulfil their innovation dreams.

Sim Wong Hoo is Singapore's poster boy for technology entrepreneurship. He started his company, Creative Technologies, in 1981 with S$6,000, six years after graduating from Ngee Ann Polytechnic with a degree in electrical and electronics engineering, and achieved mainstream success a few years later. His breakthrough product was the Sound Blaster audio card, an internal computer facilitator of input and output audio signals that for a number of years was the standard for many of IBM's personal computers. On the strength of this best-selling product, the company listed on NASDAQ in 1992 and expanded its products with a full spectrum, including speaker systems, electronic musical instruments, CD-ROM and DVD-ROM drives, MP3 players and webcams. Sim found his success by going to Silicon Valley, and developing a product that won acceptance in the important US technology market.

Popular online dating site Match.com was co-founded by Singaporean Ong Peng Tsin. Both Match.com and his second company, content management systems producer Interwoven, were seeded and exited in the USA, with the former sold to IAC/InterActiveCorp for US$50 million in 1999 and the latter sold to US-based Autonomy Corp for US$775 million in 2009. In Singapore and in other countries with less developed start-up scenes, their talents have moved to the Valley and elsewhere abroad in search of the right support to nurture their ideas. While many of the talents have returned to make contributions to the home country, the innovation processes and learning they created while overseas would not easily translate to spin-outs in Singapore, as have been done in Silicon Valley.

Back to the question: can Singapore create a vibrant entrepreneurship ecosystem ala Silicon Valley? The Silicon Valley success model can be summarized simply as:

$$\text{Innovation success} = \text{Talent} + \text{Ecosystem} + \text{Market}$$

Talent

As can been seen from the successes of Ong Peng Tsin, Sim Wong Hoo, and the technopreneurs profiled in this book, Singapore does not lack *talent*. Many scholars have taken up a popular pastime to criticize Singapore's rigid and exam-centric early education system as the key culprit for the lack of entrepreneurial talent in the city-state. I beg to differ from this stream of argument. In my interactions with many entrepreneurs as part of my advisory roles with NUS, SPRING,[3] and at SMU, I do not see a lack of innovative ideas from Singaporean youths. The only dimension where I can see improvements are needed is in the impact of their ideas. Singapore being a small city-state, many young entrepreneurs think only in terms of the home market or the ASEAN region as the target addressable market. In a fiercely competitive and rapidly globalized world where the internet enables the spread of technologies across borders, Singapore's entrepreneurs should set their sights on the world market with interim goals of regional and local market domination as a first step in business model validation. But, with the right mentoring and coaching, the key ingredient of talent in this equation should not be an issue.

Ecosystem

The *ecosystem* variable is a much harder component to crack, and Singapore is making gallant efforts in creating a healthy funding base for early-stage investments. Over the past two decades, based on my behind-the-scenes view of the local entrepreneurship scene, I have observed that many apparently promising start-up failures can be traced to a lack of a robust funding environment for early-stage technology ventures. Often, problems stemmed from a lack of experience across the entire ecosystem; too often, entrepreneurs gave away too much equity or newly-established investors put forward impossible terms. To grow the economy and to prevent another country from benefiting from the fruits of its homegrown stars, Singapore is working to develop a supportive atmosphere that is able to take big ideas to the next level.

In 2002, Singapore created a 'Seeding for Surprises' sub-committee as part of its overall economic restructuring committee effort. This sub-committee's goal was to create and cultivate a vibrant entrepreneurial ecosystem with seed funding, incubators, and mentors for young start-ups[4] and targeted the issue of how to create a sustainable capital base for innovation. Today, the fruits appear in the form of number attractive funds and schemes aimed at both keeping start-up talent in the country and enticing talent to relocate from abroad. Ong Peng Tsin returned to Singapore in 2001 and is now a critical contributor in the development of this ecosystem. He is a board member of the Singapore NRF.

NRF was established to shepherd the innovation process from knowledge creation, to dissemination and its value creation, under the guidance of the Research, Innovation & Enterprise Council (RIEC). While educational institutions and research institutes were already involved in cutting-edge research, too few had successfully turned their findings into commercially viable results. An initial investment of S$5 billion was committed to the NRF for the creation of a range of funding platforms in three main areas: for various educational institutions and research institutes, for proof-of-concept and proof-of-value grants to support the translation of innovations to applications, and in 2008 a $360 million budget for the creation of the National Framework for Innovation and Enterprise (NFIE) to advance technology innovation into the marketplace with commercialization initiatives.[5] A key strategy in NFIE is to support academic entrepreneurship and fund IHLs (institutes of higher learning, i.e. universities and polytechnics) in their incubation activities and to fund a network of business coaches, advisors, and experienced entrepreneurs as mentors to the younger generation of start-ups. This is a key success enabler in the ecosystem variable. As I asserted earlier, talent in Singapore is not lacking but with appropriate coaching and mentoring from more world-view advisors, the small 't' in talent can be amplified to a big Talent. This should be viewed as the number one priority for all who are engaged as advisors in this ecosystem.

When I was introduced to a small pre-funded start-up named Paywhere[6] in 2010, the company's original vision was to help bricks-and-mortar fashion boutique stores to establish an e-commerce storefront via a set of productivity tools. Through a series of critical questioning and feedback sessions, the three young Singaporean co-founders are now embarking on a far more ambitious strategy of being the premier social-commerce provider for the global market (teaching case included in this book). Of course, an amplified vision requires the necessary funding support to realize this dream. Without this critical platform, entrepreneurs with big visions have no choice but to relocate to other countries to seek funding if they want to validate their business model quickly.

Fortunately for Singapore, funding is not a problem at the macro level. Government agencies have various schemes launched to sponsor/grant start-up ideas. But there remains a gap between the talent and the sources of funding, which was how to translate the idea into a fundable business model and business plan. These attractive schemes have also attracted a number of incubators, accelerators, and investors to set up shop in Singapore. Since 2006, a number of incubator support schemes were launched by NRF which saw the formation of new incubators. As can been seen in the funding map for Singapore's technology ventures, there are now multiple sources for funding for the idea to seed stage.

One incubator, Jumping Frog Digital Incubators Asia (JFDI.Asia) operates a 100 days ideas-to-investment accelerator 'bootcamp'. Set up in 2010, it is modeled after Boulder, Colorado's TechStars. The two co-founders and key principals behind JFDI are Hugh Mason, a British media veteran previously with the BBC and Prembridge Partners, and Wong Meng Weng, a Singaporean who started PoBox.com and Karmasphere in Philadelphia and Silicon Valley, respectively, and returned to Singapore in 2008 for family reasons. Their first venture in 2009 was to co-found hackerspace.sg a non-profit community organization providing co-working space for self-styled 'geeks and creatives' to share ideas and start collaborative projects. Some members were proto-entrepreneurs while others simply wanted to explore what technology makes possible. The first such space in Singapore, hackerspace.sg has subsequently become the prototype for around ten similar co-working spaces of different flavors. After running a successful MDA-funded pilot program that engaged local entrepreneurs with visiting mentors from around the world, JFDI.Asia's full accelerator bootcamp program began recruiting participants in 2011. SPRING's

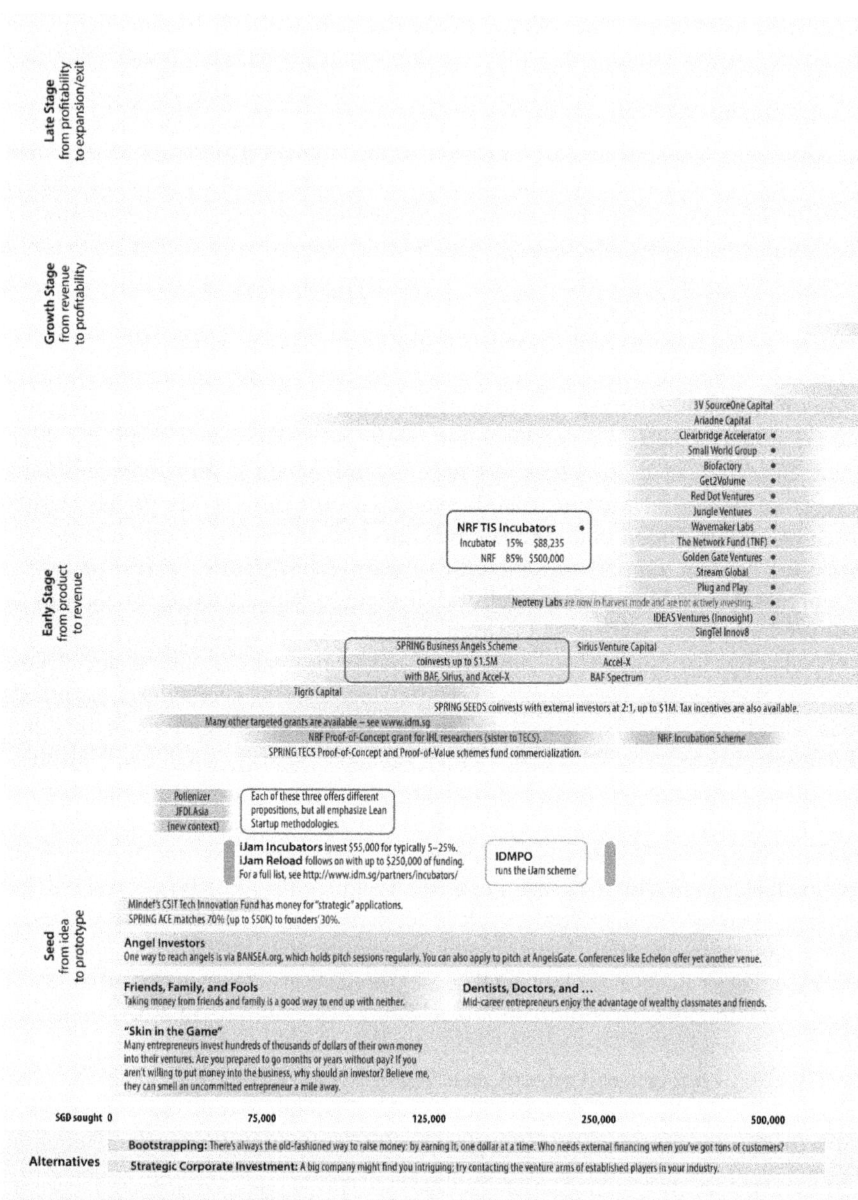

Figure III.2 Meng's Map of Money by Wong Meng Weng (available at www.mengwong.com/sg/capital/megs-20120810.pdf).

If you're playing at this level,
you probably don't need this map.
Financial institutions have private equity arms
that would be delighted to help take you public.
Intel Capital
JAFCO Asia
JAIC Asia Capital
OWW Capital Partners
Gobi Partners
Vickers Venture Partners
Vertex are the VC arm of Temasek, and appear to be more engaged with the early stage community than most funds of this size.
IDG Ventures operate out of Vietnam but claim to take an interest throughout Southeast Asia.
SBI Venture Capital
iDM Ventures (idmvc.com)
Mercatus Capital
IDA's IIPL will co-invest with VCs in Singapore IT companies.
SPRING Bridging Loan Programme matches bank loans 8:2.
iGlobe Group have been around for a while and are respected investors.
Walden International are respected investors.
Extream Ventures
Raffles Venture Partners
Ambient Sound Investments is Toivo Annus's investment vehicle.
3V SourceOne Capital
NRF Early Stage
coinvested 50:50 in VC funds
SingTel Innov8 are one of the most active seed and early stage investors in Singapore, but you have to show a strategic benefit for the SingTel Group.
Meng's Map of the Money
Source: Wong Meng Weng
Available at www.mengwong.com/sg/capital/
Placement on map is approximate.
Not all funding sources are listed;
not all listed sources are active.
Industry-specific grants may be available.
Investor appetites vary constantly.
Comments and corrections welcome at
mengwong@jfdi.asia
version 20120813
1,000,000
2,000,000
4,000,000
8,000,000
16,000,000

Incubator Development Program provided support, with SingTel Innov8[7] acting as both a cash sponsor and giving access to its network of operating companies to recruit start-ups across the region. The balance of funding for JFDI came from a network of private investors interested in participating in Singapore's next technology stars. JFDI.Asia plays at the beginning stage of entrepreneurship: in its first year it offered micro-investment of S$ 15,000 to selected teams in exchange for 5–20 per cent equity. Valuations at the start of the program were typically in the S$150,000 range, rising to S$1.5 million at the end. The market acceleration support focuses on intense mentoring but also includes physical space and in-kind technical facilities, seminars, and of course access to seed investors. JFDI sources candidates from direct online applications and from organizing events in six cities: Bangkok, Jakarta, Manila, Melbourne, New Delhi, and Singapore.

So far, JFDI is showing some early results. In 2011, the incubator completed their first batch of investee companies with 11 'graduates' from an applicant base of more than 300 start-ups around the region. Eight of the graduates have received seed-round funding offers in aggregate exceeding S$4 million. Based on this early success, JFDI is gearing up to achieve a stretch goal of 100 start-up teams by the year 2015, further adding to the vibrancy of the developing ecosystem, especially on the beginning stage of an entrepreneur and the germination of the idea. Other incubators operating at this very early stage are the university incubators where space, coaching, and various grants up to S$50,000 are available in support to take the students, alumni, and staff ideas to the seed stage. The local universities, NUS, NTU, and SMU, and polytechnics all have entrepreneurship centres offering a range of incubation support.

When ideas reach prototype or beta form, there are now a number of seed-stage investors for funding the Series A round. The first scheme launched by NRF's NFIE is the Early Stage Venture Fund (ESVF) scheme[8] where NRF took a fund-of-funds approach and invested S$10 million each into six qualified fund managers. The fund managers have the option to buy out NRF's shares within five years, giving them added incentive to maximize their returns through investments into high-potential start-ups. The fund managers from ESVF would typically invest in promising start-ups in the range of S$1–2 million in a Series A round. Some of the successes from this earlier scheme included HungryGoWhere[9] and Brandology.[10] With this momentum and recognizing that there existed a gap in the funding stage of seed-stage of sub-S$1 million, NRF launched the TIS in 2009 to jumpstart a greater number of incubators and market accelerators to help start-ups to achieve their next level of idea validation. The TIS, inspired by Israel's Technological Incubator Program, seeks to address Singapore's underdeveloped start-up scene by building a healthy ecosystem through quality mentoring, funding, and space incubation from a number of approved incubators. Basically, the scheme derisks early-stage investments for incubators. A qualified TIS incubator contributes expertise in selecting appropriate start-ups and is able to invest up to 15 per cent, or S$89,000, while NRF acts as a joint investor contributing the remaining 85 per cent up to a maximum investment of S$500,000. If an investment goes south, an incubator merely loses the initial investment, a small amount in the range of an angel investment, but because of the NRF matching equity investment, these investments now come packed with the potential scale and impact of an early-stage institutional investor. A typical TIS investment valued at S$2 million pre-money would see the start-up dilute by 22.75 per cent and a board seat. The scheme is structured as a convertible bond with the incentive to allow the incubator to buy out the NRF's S$500,000 bond at the original value with nominal interest in three years. The objective of this scheme is to encourage seed-stage investments and has the net effect of turning qualified angel investors into scalable incubators to fuel the entrepreneur ecosystem.

The TIS has also attracted Silicon Valley immigrants. Golden Gate Ventures and Wavemaker Labs are recent examples of financiers and entrepreneurs from the Valley who setup an incubator in Singapore to take part in this scheme. As typical of Singapore government's openness to take good ideas from anywhere, incubators do not need to be started or run by Singaporeans nor are they limited to investing only into Singapore citizens' start-ups. The TIS incubators just need to invest in seed-stage technology start-ups with major development (especially R&D) and operational activities in Singapore. As of mid-2012, 40 companies have received investments through the TIS with many valuations given through convertible bonds, in the S$1 – 1.5 million range, with a small number achieving S$3 – 4 million pre-money valuations. Of this number, more than half have made positive progress in their venture growth: 17 are cashflow positive, two have bought out NRF shares, and seven have moved to a higher round of funding.[11]

The scheme is attractive to both incubators and would-be start-up fund seekers, but there is one cautionary note. This form of funding is not common for seed-stage investments, and many of the start-ups would expect to need to raise more capital for their continued development and growth. Unless the next round is raised in Singapore where they would be familiar with this scheme, most fund managers would not want to invest in the next stage with convertible bonds outstanding on the company. Very likely the next-stage investors would require the seed-round bond holders (NRF and the TIS incubator) to convert the bonds as a condition precedent to the investment. It is expected that the TIS incubator would have to fulfil the conversion on his own and on the NRF bond, given that the purpose of this scheme is to de-risk early investments and not for government to hold shares in start-ups. The time when the TIS incubator is prepared to commit the full additional S$500,000 may not be the same as the start-up's need to raise more capital for its growth. If the tales in Part I and Part II of this book offer any insights from the past, entrepreneurs who raise money from TIS incubators should work closely with their incubators/investors to ensure that there are no surprises on the path of growth. More so than ever, the seed-round incubator has the power to determine the timing (and, maybe, the terms) of the next round. For all matters associated with shareholding and capital, the incubator is more of a 'co-founder' than an 'investor' of the start-up during the seed-round stage, and the founding team needs to keep close engagement with their TIS incubator, beyond the standard board meetings.

Nevertheless, what is interesting about the TIS is not only its effective support for seed investments, but that it has become a platform for entrepreneurs of yesteryear to continue their engagement in technology venture creation as a mentor, investor and coach, one of the critical ecosystem support components of Silicon Valley. In 2009, I conducted a number of interviews with local technopreneurs for my PhD thesis. As I was interested in the topic of professional experience and how it influences opportunity recognition and the new venture-creation process, the selection of the interviewees drew from technical professionals that had spun out of IT MNCs. It speaks to the power of how the ecosystem has developed in three short years when a large sub-group of the 18 technopreneurs I conducted in-depth interviews with then became key players in this early-stage investment community. Chak Koon Soon and Ronnie Wee are active as TIS incubators, and Ng Fook Sun and I are angel partners to Leslie's Red Dot Ventures.

Chak Kong Soon and his firm Stream Global was one of seven incubators[12] selected in TIS's first phase. In 2012, a second phase saw eight more incubators[13] selected. In this phase, Leslie Loh's Red Dot Ventures and Ronnie Wee's IncuVest were selected. Leslie and Ronnie had already been active participants in the earlier stage of this ecosystem as managing partners of NRF co-funded Extreme Ventures and SPRING-supported Sirius Angel Fund, respectively. Ronnie's Sirius Angel Fund invested in a number of early-stage tech companies, the most notable being Reebonz, a successful Singapore-based Asian flash sale e-commerce site, which raised US$25 million from Intel Capital in its latest round. There were a number of other good exits and not one investment actually lost money. Sirius Angel Fund was closed off in 2011 with a successful track record clearly in place. A sterling winning track record made it a natural for his IncuVest to secure TIS.

Winning TIS incubator status was a natural extension and further deepening of their engagement in the developing ecosystem.

Former entrepreneurs are in a unique position to leverage their personal experiences. Investors from purely financial backgrounds have not gone through the joys and struggles of building a business, and may not know the nuances of what to ask today's start-ups. To encourage his investees, Chak encourages them to 'cut the bull' and be honest with their struggles. From being in their shoes, he knows new entrepreneurs want to avoid sounding stupid with unfocused ideas in front of their investors. But without honest dialogue, he and his Stream team, which consists of former entrepreneurs, fund managers, and consultants, can contribute a well-rounded perspective of where the gaps are and how they can be overcome to execute a business successfully. One area where start-ups often need assistance is in raising their next round of funding. Chak's personal preference is to give investees a low valuation rather than taking a callback, which he sees as a de-motivator. Most investors will ask for four times liquidity but he sees 1.5 or two times to be fair. While this may seem low, as a government incubator, their focus is on growing companies and building trust in the nature of Singapore's entrepreneurship ecosystem.

When asked about his personal investment philosophy for strengthening entrepreneurship in Singapore, Ronnie responded 'entrepreneur, entrepreneur, entrepreneur.' His past as an entrepreneur influences him to invest big in promising entrepreneurs to build the ecosystem. While a good idea is important, he values an entrepreneur's character more highly. The nature of early-stage ideas sees them morph on a regular basis, but with the right investor/investee relationship adjustments can take an idea to the next level. Ronnie hopes he and the IncuVest team will be able to form such relationships and execute big ideas. Their combined decades of experience wove a web that they can now tap into to quickly validate ideas; it is within their ability to call up telcos in Indonesia or Thailand to run pilots. This kind of geographic reach, in addition to innate risk-taking entrepreneurial tendencies, makes them more open to investing in big ideas. While various historical and cultural factors challenge ambitions to build a 'Google' in southeast Asia, Ronnie remains hopeful that ex-entrepreneurs like himself, who invest in a portfolio of companies with different speeds and market positions, will be able to nurture winners who will get to that level.

With a rich tapestry of experience building one of Singapore's most successful software companies, Leslie has a trove of lessons to share with today's aspiring entrepreneurs. The key factor that distinguishes his journey is a willingness to take risks. Too many entrepreneurs are unwilling to take big risks. Leslie says. 'The chances of you making it are 5 per cent, 10 per cent, 2 per cent, but as an entrepreneur, you have already taken the main risk so you might as will bet.' Companies do not defy the odds and become the exception by being conservative. 'If you might die anyway, make sure you die gloriously!' he says. However, one way to avoid sudden death is to start on the right foot. Leslie says he has seen too many spin-outs fail due to an even distribution of power. Rather than have five co-founders with 20 per cent each, there needs to be one clear captain. A leader who will make the tough calls when necessary is essential, and share distribution needs to reflect this. A few years down the road, it is likely that two or so of the original founders will leave and the leader will still own only 20 per cent of the company. Without skin in the game it will become difficult to progress. It is important for leaders to be player-coaches and in tune with the team and with the realities of the field. Being genuinely interested is key for success. 'If you are arrogant or not good with people, forget about being an entrepreneur. People will go with you because they like you', says Leslie. However, perhaps the secret to System Access's success is their willingness to gamble, to continuously innovate. In an era of rapidly changing technology, the company took chances on the PC and on Oracle; luckily, they were the right choices.

Fook Sun believes his value as an investor lies in his ability to link big ideas with a practical plan for execution. As a former entrepreneur, he feels a sense of comradeship with his investees as he is able to relate to and share the excitement of building a business. This makes it easy for him to establish genuine trust with the right start-ups. With trust being the foundation of a strong investor/investee relationship, he is able to aid them effectively in translating their idea to set goals and milestones in the early stages. At the same time, he is able respectfully to share personal lessons from his journey, especially how he excelled by pivoting towards marketplace needs. A disconnected relationship with investors is a frequently heard complaint among start-ups, but Fook Sun thrives on relationship-building. With genuine support and guidance, he is a member of a new generation of entrepreneurs-turned-investors that are nurturing Singapore's aspiring start-ups towards success. The ability to find such support contributes to a robust culture that will continue to both retain and draw in twenty-first century entrepreneurs.

With healthy and vibrant early-stage investment funds sprouting up, the ecosystem is developing rapidly to attract many new technology start-ups. In SMU, young start-ups such as Innova and Rainmaker Labs, have already received term sheets from TIS incubators. At NUS, many start-ups, including Plunify, 2359, Stream media are funded by TIS incubators or by Singtel's newly formed Innov8 funding arm. NUS, Singapore's Media Development Authority (MDA), and Innov8 joined forces to launch a space incubation program at Blk71, a set of converted industrial blocks with funky colors across from INSEAD's Asia campus, near Fusionpolis. MDA has a funding scheme named iJam[14] which, like SPRING's ACE scheme, grants an initial S$50,000 but comes with mentoring and incubation support. Upon sequent successful investments from incubators/investors, iJam recipients are eligible to receive matching one-for-one grants for equity investments up to S$100,000. None of these funding schemes guarantee success but there is now an undeniable buzz in the hyperactive technopreneur ecosystem in Singapore. There are many inspirational stories floating around Singapore's entrepreneurship ecosystem. Today's generation can aspire to achieve Sim Wong Hoo's or Ong Peng Tsin's success, but with one difference – to be able to launch their new venture in Singapore and to have access to experienced entrepreneurs as mentors and advisors.

The goal is that these funding initiatives will turn Singapore into a global hub for innovation. No longer thought of as dull or emerging, Singapore is one of the Asia-Pacific region's brightest stars. A strong economy and modern infrastructure is luring entrepreneurs keen to make a name for their ideas in an environment that is less competitive, and closer to less tapped emerging markets, than the Silicon Valley behemoth. Of course, there remains a missing gap in the funding ecosystem – that of Series B funding for technology ventures, a quantum in the S$2–5 million range. As described earlier and made obvious from the funding map, many of the activities are concentrated in the lower end of the funding quantum. Oddly, there is funding at the larger scale via Temasek Holdings' wholly-owned subsidiary Heliconia Capital Management Pte. Ltd. Again taking the fund of fund approach, the government will co-fund up to S$250 million to match private funds into later-stage companies with revenue up to S$100 million.[15] Without further development in the higher quantum, Singapore may still see an exodus of young ventures needing to move to more robust ecosystem cities to seek more funding support and expand their technological influence.

Today's young entrepreneurs are increasingly looking for alternatives to starting their business in Silicon Valley. The Valley remains the world's most robust entrepreneurship ecosystem, but it is not always an appropriate location to headquarter a company. High prices and heavy competition influence some firms to seek growth elsewhere. Singapore is increasingly being chosen as a top destination with its growing reputation as an innovative and global city. The 2012 edition of the Global Innovation Index, published by leading business school INSEAD, examined the innovation ecosystems of 141 countries and ranked them based on factors such as human capital and research, knowledge, and technology output, and business sophistication. For the second year, Singapore took Asia's top spot and third overall, with Hong Kong as the only other Asian country to break into the top 20 in eighth place. Singapore was ranked number one in the world for the innovation input sub-index, which comprises institutions (political/regulatory/business environments), human capital (education, R&D), infrastructure (especially ICT), market sophistication (credit, investment, trade, and competition), and business sophistication (knowledge workers, innovation linkages, knowledge absorption).

With Singapore developing a strong reputation of being an innovation hub, many top companies are lured to set up operations locally, thereby adding to the vibrancy in the entrepreneurship sector. In addition, NRF developed a Global Entrepreneurial Executives scheme to attract high-growth, venture-funded, high-tech start-ups with experienced entrepreneurs to relocate and set up core operations in Singapore. Through this scheme, NRF will invest up to US$3 million in matching funds, again through convertible bonds, through IIPL (Infocomm Investments Pte. Ltd) and EDBI (Economic Development Board Investments), two Singaporean VC firms set up with initial funding from government agencies. IDA has an iCentre program that provides developmental support for start-ups to establish new engineering centres in Singapore. With over half of their 50 million users residing in southeast Asia, it made sense for Silicon Valley founded and funded global mobile community Mig33 to relocate their operations to Singapore. Their inexpensive virtual gifts, chat rooms, and games appealed to fun-seeking non-smartphone users. In southeast Asia, where a high percentage of the population has at least a basic mobile phone, all it takes to run Mig33 is a java-based phone. Their popularity was growing at an exponential rate. For IDA, their relocation builds on the quality of entrepreneurship and engineering efforts in Singapore. Today, ease of doing business, robust R&D funding, and a strong education system are but three commonly cited reasons why Singapore is an excellent place to conduct business and why foreign investment continues to increase.

Market

Of the last variable, *market*, success may elude Singapore's young start-ups. In this regard, Silicon Valley or any US-based start-up has an unmatched social advantage – a large market full of tech-savvy and willing early adopters. A start-up in the USA can get the ecosystem advantage from being located in the valley, but a start-up in other states can also gain momentum via rapid customer adoption as the USA is a vast early adopter market with largely homogenous buying patterns across its 50 states. A Singapore tech start-up must overcome a number of important hurdles, the most important being the difficulty to secure the first reference accounts, and this creates substantial barriers to any young venture in its early journey to validate its business model. Singapore's home market weaknesses can be summarized as: small, conservative, and risk-adverse for the important enterprise sector and, to a slightly lesser degree, the consumer adopter sector. The immediately surrounding region offers a collection of very heterogeneous and diverse markets, making individual market penetration difficult and time-consuming. A small home market means even after a start-up gains traction at its home base this does not guarantee any scalable revenue or sales to sustain its expansion. This issue is compounded by the mostly conservative corporate Singapore. Singapore's GLCs (government-linked corporations) are known among the start-up community as being conservative and resisting to adopting start-up solutions. They prefer to buy technological solutions from established giants such as IBM, HP, Oracle, etc. This is understandable and predictable. CIOs do not get fired for buying technologies from established giants; but if they take a chance on a young, unproven technology, the risk would be high for the corporate executives. In the USA, however, executives are not necessarily hampered by this *kia-si* (local Singaporean slang meaning afraid to die) mentality and would be open to an innovative offering if it means a leap in competitive advantage for the large company. Once again, this is the social advantage enjoyed by the US start-ups. Even corporate America is in the cheering line for entrepreneurs in the USA. For the few start-ups who manage to sell into local corporate base Singapore, they win very many sales dollars for their effort as large companies here view purchasing from a start-up as a 'gift' from a large reference account and are extraordinary stingy when imparting sales to the weaker firm. Leslie Loh never once sold to a local enterprise, and grew his entire multi-million System Access business from selling to overseas customers. Ronnie Wee lamented that he was unable to charge market rates for his PlaNET's solutions services after he left Accenture even though he was the same consultant delivering the services. Star+Globe won its hard-earned revenue almost entirely from overseas customers. Danny Wilson would scoff at the idea of selling locally and never even tried. It is this last variable, I believe, which could become Singapore's Achilles' heel in creating a vibrant entrepreneurship and innovative enterprises economy. Until the market variable can be a viable option, young technology start-ups will still need to venture overseas rather early in their journey, while they are still immature and ill-equipped to take on dilution of resources and attentions to conquer overseas markets. The one bright sector for Singapore is the aggressive adoption of smart phones (especially the Apple iPhone) by its consumers, making it an ideal test market for innovative apps for the Asian market. A winning app can go global instantly, thanks to Apple's iTunes and other app-download platforms. Perhaps with the right *Talent*, the now buzzing *Ecosystem*, and this new consumer-driven *Marketplace*, we can look forward to some successes in this sector.

With the talent and the ecosystem variables in place, this still is not a guarantee of success but there is now an undeniable and active technopreneur ecosystem in Singapore. There are many inspirational stories floating around Singapore's entrepreneurship ecosystem. Today's generation can aspire to achieve Ong Peng Tsin's success, but with one difference – to be able to launch their new venture in Singapore and to have access to experienced entrepreneurs as mentors and advisors. Those featured in this book along with the story of Sim Wong Hoo, Creative Technology's CEO and Singapore's unofficial poster boy for entrepreneurship, are proof the infrastructure is developing. Seeds continue to be sown, with some sprouts visible, but with continued fertilization in the form of varied schemes hopefully in the form of market access support schemes and local corporate enterprises' innovative cultural development, we await a rich harvest in the future.

Conclusion: Engineers, marketers, and futurists

The stories of many other entrepreneurs have often been told along the lines of 'local boy makes good' narratives. I wrote this book so the younger generation can understand the deep challenges of technology entrepreneurship. Too often, the media only writes about the successes, and any challenges are couched in a positive, 'here's-what-I-learnt', post-success type of reflection. Failures actually create deeper lessons, but the Asian save-face mentality does not encourage many of these stories to be published. After reviewing the more than 100 cases at my disposal, I picked these stories to highlight the theme – which is that technology entrepreneurship is about making strategic choices at the inception of the start-up and resolving the dilemmas confronted in the day-to-day running of the business. Often, these dilemmas stem from the early choice on how to fund the business.

This book does not focus only on successful entrepreneurs – in fact, the three stories at the start frame the difficulty in starting a technology business and the choices entrepreneurs must resolve regarding when to raise money and who to raise money from to grow their business in a fiercely competitive technology race. Entrepreneurs must take money from the right investors who share their vision of building a technology company and who also understand their business.

Taking money has a cost: All technology entrepreneurs should raise money with this knowledge in mind. But on the other hand, rule number 1 of entrepreneurship is: do not run out of money. This is the tough choice. The lessons from the technopreneurs from the mid-1990s are still relevant to today's new start-ups. Hopefully, younger entrepreneurs today can draw upon the lessons from my generation and make the right choices in who they want to partner with. Taking venture capital money does come with an important non-financial advantage and it is one of the key success factors for Silicon Valley's success. Venture money gives your venture the necessary resources to accelerate your experiment (also known as your business model).

Technology start-ups need to fail/succeed fast: The decisions entrepreneurs make early on sets the venture onto a certain path. In the second group of narratives, the technopreneurs did not start off as a technology company. Finesse Alliance started off making banking software in the systems integration in banking and financial services realm. By merging with companies to strengthen its technology portfolio, it was able to exit the company. PlaNET Technology

Solutions and System Access both started out as turnkey software developers, and quickly pivoted to focus on specific product areas where they could carve out a niche and grow the company along with the growth with their chosen sector. By selling solutions, the companies developed their product vision only after very deep engagements with customers to understand their problems. The common theme from this group of entrepreneurs is …

You let the market tell you what to build: With this strategy, you do not need to raise upfront venture capital. In fact, it would be difficult for investors to invest in you since you would be unable to draw up a realistic business plan to chart your course. You just need to have good antennae and a good engineering team and trust in market growth. If you are earning customer revenue in a scalable fashion, you will become a profitable business, barring any financial errors. Wait for your quantum leap opportunity, because the market works in cycles. Keep in mind …

A rising tide lifts all boats: You do not have to be the fastest boat, you just have to outlast everyone else when the tides recede, and remain as one of the strongest ones when the tides return. To do this, you have to be able to persevere with your vision without external intervention. System@Work focused, from its beginning, on mobile payment technologies, and was able to keep to its vision even through challenging times because they did not raise external capital. All eventually received buy-out offers from international companies wanting to gain foothold in Asia's growing market. They had to trade off difficult bootstrapping early days and defer their product development goals until after they gained sufficient financial strength from customer revenue to keep their autonomy. To the casual reader, it may seem like the narratives in this book are trying to convince technopreneurs to eschew external capital and grow the business organically. Certainly, the path chosen by the successful entrepreneurs in this book is a proven model. The problem with the organic growth model is that, by definition, you are working on near-term problems. This model works for ventures with an engineer at its helm. You should choose this path if having control of the destiny of your company and solving problems with engineering solutions is your forte.

Some entrepreneurs dream of being able to create the next Google. To do this, you have to be able to project future needs. In addition to being a good engineer, you also need to be a marketer and a futurist. Steve Jobs fits that description – actually he was probably more a marketer and a futurist than an engineer. In contrast, Singapore's Sim Wong Hoo was the engineer and futurist. He found success by creating a technology application that foresaw a consumer demand for the ability to create, hear, and edit quality sounds on a digital platform, and by being the first to market an engineered product based on the technology.

Futurists need to have such a strong belief in their vision of the future that they scoff at the need to consult with market research and analytics. You need to believe in this future vision even with the lack of data to support you. Steve Jobs famously said 'It is not the customer's job to know what they want.' Sim Wong Hoo took a chance on his vision. He created a new industry based on anticipation of future demand and lifted the ceiling on preconceived notions of what is possible. That Creative was able to become a household name in a then up-and-coming industry is inspiration for many local start-ups that are seeking to pioneer or create emerging industries today. A Google, Apple, or Creative needs visionary venture capitalists to support them. In Creative's case, they found a believer in Walden International, a global venture capital firm started by Lip-Bu Tan. However, visionary venture capital firms willing to gamble multi-million dollar funds on a technology start-up are few and far between in Singapore and many Asian countries. This is the one huge gaping hole in many countries' innovation growth strategies, and Singapore has set out to address this gap.

Singapore has her fits and starts in its ambition to emulate the Silicon Valley model. In the late 1990s, various branches of the government were tasked to bring a new economic development program, the Technopreneurship21[1] (T21), to reality. As part of the T21 initiative, Singapore put aside a US$1 billion Technology Investment Fund (under TIF Ventures) to boost investments in technology start-ups, in a fund-of-funds approach intended to attract foreign VCs to set up shop in Singapore and bring their investment expertise and transfer their knowledge and train up new teams of venture capital professionals.

Sadly, those efforts did not create the expected outcome in igniting the local entrepreneurship scene. Many of the big-name venture capital firms did set up shop in Singapore as the funding incentives were attractive, but then promptly decamped to China where they invested in what are now household names – Baidu, Alibaba, Sina, etc. In fact, Venture TDF no longer participates actively in Singapore's technopreneur community. Their investment executives are focused mainly on China-based new ventures. This is understandable as finance people do chase after fast-growing markets and are motivated by high return on investments. To build technology start-ups in a small market is incredibly hard work, even more difficult than to build comparable companies in a large market, and yet with much smaller returns – so the natural question investors would ask is, why bother? VCs are not the right partners to take on the hard work of anchoring in Singapore and pulling up their shirt sleeves to work with young entrepreneurs to build up the value. Fortunately, there are now enough technology entrepreneurs who have found exits and are willing to pay it forward with incubating the younger generation of start-ups and Singapore's government schemes are crucial in creating the platform for the building of impactful, indigenous technology start-ups. The new NRF funding schemes now have more appropriate measures and schemes to ensure that a good mixture of local and foreign mentors set up shop to incubate and nurture start-ups, and not just to impart their investment expertise or just to take government incentives to build a financial business.

Singapore has the ambition to cultivate local talent and build the local entrepreneurship ecosystem. The Little Red Dot is home to many promising ideas that could thrive with the right support. With the right combination of Talent + Ecosystem + Market, I predict that we will see at least 10 to 20 successful exits, and one Asian 'Google', before the end of this decade.

I plan to be an investor/coach in those start-ups in Singapore.

Part IV

Paying it forward

Preparing the new generation of start-ups for success

Teaching case study 1: Paywhere

Dickson Gregory (CEO), Vincent Lau (CTO) and Damian Chow (CFO) were convinced that they could execute a solution that would create value for those in the e-commerce realm. Their multidisciplinary backgrounds and previous experience in the e-commerce industry provided insight into problems for merchants in creating an e-commerce shop. One year after the soft launch of their revolutionary product TackThis!, Paywhere garnered many strategic partnerships. However, despite an almost exclusive devotion to fundraising, Paywhere is still undecided on who should fund their seed round, jeopardizing their first-to-market advantage. The story of Paywhere and their journey in fundraising provides insight into the realities of launching an entrepreneurial venture juxtaposed to an immature entrepreneur financial ecosystem.

Introduction

Internationally recognized as a progressive and forward-thinking nation, Singapore has accomplished a degree of development unlike any other country in the world today. In one generation, the city-state transformed from a developing to developed country, earning many accolades in many world rankings.[1] The country's forward-thinking policy-makers continue to focus on how to develop the nation to succeed in a future where high technology manufacturing is rapidly shifting to newly emerging economies. One skill that many believe requires further development is creativity and innovation.[2] With her strong culture of entrepreneurship, it is no coincidence that many of the world's most innovative ideas – such as Facebook and Google – came from the USA. Singapore is hoping to follow a similar path and has embraced the Silicon Valley model of creating a vibrant ecosystem for nurturing entrepreneurs and new venture creation activities. A plethora of grants, incubation support, and research funding to grow this entrepreneurship ecosystem has been unleashed since 2009. One local firm that has benefited from new government initiatives is Paywhere.

Figure IV.1 From left: Damian Chow, Vincent Lau, Dickson Qirui Choon Gregory.

How it all began

In 2009, Dickson was finishing his final year at NUS and pondering his next step. He was certain that he wanted to start an entrepreneurial venture rather than take the conventional Singaporean routes of pursuing graduate studies or climbing the corporate ladder. For him, the life experience he would gain as an entrepreneur would be invaluable. It also just sounded like fun.

Dickson's former classmates from his junior college days, Vincent and Damian, concurred. At the time, Vincent, Damian, and a fourth founder, Gerard, were helming Aeturnus, an e-commerce firm they founded when they were in secondary school. The company was originally founded as an outlet for them to play around with web design, but over time they progressed into programming, content management systems, mobile websites, and hardware. The business was kept operational as the founders pursued their university studies, which enabled them to build confidence in their entrepreneurial skills. While business was steady, the opportunity to pursue a new innovative technology was appealing. That they would be able to start a new firm with an old friend was the icing on the cake. Looking back Dickson says, 'What really enticed us was that we would be doing things together. [We felt it was fun] to work towards a common dream for our company'.

Go-getters by nature, they had confidence in their grasp of the field and chose to act by creating Paywhere in May 2010, a firm focused on e-commerce solutions, rather than over-analyzing the situation. They promptly entered the incubation program at NUS (see appendix for support offered by NUS's entrepreneurship incubation program (NEI)). Looking back, Dickson jokes: 'So we just started venturing... sometimes we maybe don't think too much'.

Jokes aside, the founders' fun-loving and laidback personas belie an intense focus. They are dedicated to transforming the e-commerce industry with Paywhere, realizing that time is a precious commodity in the technology marketplace.

The importance of being first to market

A few years earlier, Vincent had conceptualized an e-reader. He created a prototype and sought a partnership with the Singaporean Agency for Science, Technology and Research (A*STAR) in 2006. While they provided him with some initial funding, third-party commercial investors were difficult to secure because they wanted to see more than a prototype, and A*STAR's policy required that another investor be committed to the project before additional funding can be unleashed. In Singapore, government agencies typically prefer to work in tandem with independent third-party investor(s), and match investment dollar for dollar. Such a relationship appears to exist to check and balance the funding decision. Vincent muses, 'Maybe it's a Singapore thing, but for the government, they do not really want to fund the money if there is no other investor willing to join them in the funding'.

This lack of synergy produced what he calls the 'zigzag' effect, where one party would not invest unless the other did. Late 2007, Amazon announced the Kindle and shipped internationally in late 2009. An opportunity was lost. What this experience taught him was the importance of action. While he believed in the farsighted vision of the e-reader, he knew that he was not the only one with great ideas. Becoming first to market is essential. It is always the first mover who gets it; whoever does it first *will clearly be the one who has the last laugh.*

The lost opportunity was bittersweet as he lost out on market share but was able to gain insight into navigating Singapore's entrepreneurial sector. With Paywhere, he was determined to push hard for funding in order to not let another chance slip through his fingers.

Conceptualizing the next big idea

In 2009, the hot topic surrounded issues facing e-commerce. With their combined multidisciplinary background, the trio was convinced that they could work well as a team and create value for those in this space. What stood out for them was the potential to capitalize on the rapidly growing number of social network users. They noticed that many small businesses were creating Facebook pages and blogs to draw interest and attention to their online or bricks-and-mortar stores. Lack of understanding on how to set up an online shop was a key barrier to e-commerce. After conversations with their peers, many of whom also fit within their target demographic, the founders felt like there was a market demand that was not being met; inadequate solutions existed within the current system paving the way for a new market to be created. If businesses were using social media pages to promote their online shops, customers would have leave the page and connect to another website creating an unnecessary extra step. What if, they thought, potential shoppers could shop for items and checkout while remaining on, for example, Facebook or Blogger? Would bricks-and-mortar stores with social media pages be enticed to create online shops if the process was simplified?

Originally, Paywhere wanted to establish themselves in the e-commerce industry as a key player in helping bricks-and-mortar merchants enter the e-commerce realm with their easy-to-use toolkit on popular social media platforms. However, their mentor[3] assigned by NEI challenged them to dream bigger. What if, the mentor questioned, Paywhere could take their idea and use it to enable everyone to be an e-seller? If they stuck with their original plan, their dream would be a service-based, medium-risk, cashflow positive idea. Businesses would pay them to create an e-commerce site and their growth would be steady. They might be able to raise one small seed round but the mentor warned that it could be difficult to raise even this seed round because such a cashflow positive business model would be an 'income-replacement' business venture and investors have difficulty realizing a financial exit. However, enabling anyone to be an e-seller would result in a 'sky's the limit' growth scenario, and if pitched properly, and funded adequately, had the potential to develop into the next big idea.

With a high number of tech-savvy Singaporeans already engaged in social media, leveraging social media to reach a wide audience was an essential strategy, so the mentor's challenge resonated with them. They expanded upon their original idea and created TackThis! as the platform to enable social commerce for anyone, not just for retailers, and embarked on their journey to create the next big thing.

Government initiatives

To get their idea off the ground, Paywhere applied for funding under the SPRING Young Entrepreneurs Scheme (YES!) and was awarded the grant in September 2010.[4] Having received the maximum grant amount of S$50,000, and self-funded the required bootstrap capital of S$12,500, the founders were able to launch Paywhere and all three founders committed to Paywhere full-time. Vincent and Damian resigned from their respective full-time jobs and joined Dickson in Paywhere.

Initial testing

From the outset, it was clear that their idea resonated with existing businesses. With limited marketing effort, 100 businesses from Singapore, Malaysia, Spain, Slovakia, and the USA requested to be part of the initial test group. They ranged from established e-commerce businesses with over 100,000 sales transactions per month to fresh start-ups funded by SPRING Singapore. Of these, 30 were selected with the requirement that they work alongside Paywhere in improving the prototype. Initial responses were extremely positive with local blogshop Hope Alethia raving, 'The versatility of this widget makes it a very useful and viable

TackThis!

TackThis! is an online platform for users to create and 'tack' a shop widget to their Facebook/blog/website within minutes and allow their customers to complete their sales transaction immediately with a seamless social shopping experience. With one synchronized inventory system, anyone can be an e-seller and easily manage all their shops on the different sales channels. Most importantly, TackThis! contains an integrated payment system via partnership with PayPal and other payment gateways, so the process of creating an online shop could be as easy as sharing a YouTube video with friends. Because of features built into TackThis!, fans of the shop could embed their favorite online store(s) to their own social media page thereby generating some commission for themselves while effortlessly engaging their social network. By simply logging on to social networking sites as per usual, customers would be reminded of the shop's presence and could purchase items directly, while their interest remained fresh, without ever leaving the webpage. TackThis! is available as a freemium product. A basic package is available free of charge, and a subscription fee is charged for the full-feature product version. Paywhere would also earn a transaction fee for every item sold on the TackThis! platform.

tool and I foresee that as more shops adopt this widget, blogshopping will be made extremely convenient for the customer as well as the owner'.

Traditionally, small businesses are constantly at risk of being swallowed by big box corporations. The concept behind TackThis! is unique as it not only supports small businesses in expanding their reach, but also targets a niche area that larger corporations have yet to address: capitalizing on raw interest and social media.

Growing entrepreneurship in Singapore

While there are a rising number of government and educational initiatives dedicated to supporting emerging Singaporean start-ups, the competitive nature of Singapore's education system has often produced students who dislike risk-taking and fear of failure. There is a growing realization that developing these characteristics among the populace is essential if Singapore aims to promote a spirit of entrepreneurship. Over the past few years, a number of government grants have been established to help new start-ups get a foothold in the global marketplace. However, critics lament that the grants blow more smoke than fire. In Singapore, founding a start-up is often viewed as the easy part. Reflecting on his experience, Vincent notes:

> The government has made it very easy to start a company, and has simplified the process. I have friends from China who say that things are actually much more difficult elsewhere, and they actually even like coming over here to set up a business. But to actually get the start-up to the next stage, is actually not really conducive over here. Not like in the U.S. and everything, from what I've heard. Funding is difficult, getting investors is even more difficult, and getting, really convincing people on what you're doing is a hell of a lot of trouble.

One of the biggest challenges on starting a company in Singapore is the lack of funding for 'ideas' or 'prototypes', unlike the investment ecosystem in hotbeds such as Silicon Valley or Route 128 in Boston, Massachusetts. Investors, even government-supported technology incubators, are themselves products of Singapore's meritocracy education system where you must prove yourself before you are rewarded. As a general practice, Singapore's investors do not invest in start-ups with only product ideas; they withhold their investment commitment until they see at least a late alpha or beta launch of the proposed innovation. However, there was hope. The assigned NUS mentor became their angel investor, and injected capital into the company when their SPRING YES! grant funding was exhausted; the capital enabled the young fledging firm to complete the minimum feature set of TackThis! for a beta-launch in September 2011. Furthermore, the company was able to partner with southeast Asia's leading blogshop platform LiveJournal to power their users' e-commerce transactions with TackThis! With these early achievements, the company embarked on raising their first serious seed round of S$500,000.

Finding the financial support for the next big thing

By year-end 2011, Paywhere was in the midst of closing their seed round of venture funding. Making it to this step was a big accomplishment for the Paywhere founders as they had gone from alpha, in January, to launching beta, in September of that same year. With a clear vision of how they intended to run the business and by coming out of a successful beta stage, they were able to meet with a number of investors. Narrowing the options down to four, they awaited offers and ultimately had to choose between offers from an incubator, a VC, and two corporate investors.

Deciding on an option was not easy for each would enable them to pursue their future via different paths. Investor money, they knew, always came with certain strings attached. The investors would have certain opinions as to how they should run their business based on the rate of expected returns. Because Paywhere had at this point developed a focused vision, it was easier for them to narrow down their options.

With an incubator, they would receive the support and mentorship they desired but many did not have the means to follow on. While they could provide seed funding, they would not be able to provide Series A or B funding which was more in line with Paywhere's vision of growth. While many start-ups would go with VC funding for their seed round with expectations of follow-on rounds, Paywhere decided against this option based on differing visions of their future growth. Firstly, Singapore's VCs tend to invest at Series A or B, and rarely at the seed round. In addition, as their main intention is to capitalize on returns with quick successes, they would prefer an exit within five years. Dickson predicts that should they not reach their targets in five years, the VCs would seek to exit. A potential dilemma occurs if the VC wants to exit at an importune time for the start-up. Of course, the valuation offered them was also of concern but the founders maintained that valuation was not the most important consideration. Often, the benefits of investments are not clear upfront but it is important that start-ups think about how an investment today will play out over the upcoming years.

Paywhere considered potential corporate investors. They were approached by a large regional publicly-listed media company in the business of online directories, which expressed an interest to make a strategic investment. This corporate investor would enable them to have an accelerated path towards regional expansion. Additionally, armed with SPRING's endorsement, they were introduced to a large and established Singapore media company. This potential corporate investor would be able to help Paywhere with branding and marketing, and offer business advice in addition to helping them secure and blanket Singapore's market. A nationwide social commerce skills competition is but one initiative that they hoped to carry out in 2012 and the investor already had the relevant departments that could help them carry out a successful campaign. Both corporate investors offered term sheets; however, because of the potential of Paywhere being able to disrupt their existing business, both investors requested an option to purchase majority ownership in the company.

In addition to the two corporate investor term sheets, Paywhere also received two offers from incubator investment funds supported by Singapore's NRF co-investment schemes.[5] Both incubators' managers are seasoned business executives, and both have a reputation for spending the necessary time, attention, and focus in nurturing their investee companies through the early stages. One incubator was concerned about Paywhere's cash-operating model and expressed his preference to keep the burn rate very low. Vincent, the CTO, was concerned that this would hamper the company from maintaining their technological lead and erode their first-to-market advantage, as Paywhere is in a fiercely competitive and innovative space with many start-ups vying to be the next big idea. Another incubator promised to be a very fast-moving investor and capable of closing the round quickly, but he was very fixated on revenue generation as the most legitimate form of validation, and requested an additional term, known as performance forfeiture, that would dilute the founders' shares if the company missed their revenue target. As the company is targeting to be the number one platform for social commerce, this goal may require the founders to pivot the business model where revenue generation is secondary to gaining user registrations, especially in the social media technology space where freemium is the operating practice. The performance forfeiture would severely penalize the founders should there be a need to pivot, which compromises the business plan revenue targets.

Who should they choose as their seed-round investor?

The founders now have a dilemma – whose term sheet should they accept? The term sheets vary in valuations to a minor degree so the post-money dilution among the four was not a determining factor. None of the term sheets is ideal, yet time is running out and they need the capital to take the company to the next stage of development. Knowing that the first investor occupies a key board seat and holds great decision-making power, this is a critical decision which can alter their fate.

Teaching case study 2: Founders' dilemma: The Gozo/around! case

Xu Da Xiang (CEO) and Quek Shu Yang (CTO) were convinced that they could execute a solution that would create value for merchants interested in offering value via the mobile space. Their multidisciplinary backgrounds and previous experience in Silicon Valley provided insight into problems for merchants in attracting and retaining customers. Within the first year after the soft launch of their revolutionary product around!, Gozo Labs garnered many grants and investment proposals. However, despite winning every award proposal they pitched, the two founders found themselves with a rapidly diminishing cash position and the investment funding at risk for completion. Meanwhile, competitors are entering the same space as around! and winning marketshare from Gozo, eroding their first-to-market advantage. The story of Gozo and their journey in fundraising provides insight into the realities of launching an entrepreneurial venture, balancing the needs of the investors and the dreams of the founders.

Introduction

As discussed in the previous case study, in just one generation Singapore has transformed from a developing to developed country, earning many accolades in many world rankings.[1] Singapore was ranked third from a list of 141 countries in the Global Innovation Index 2012 published by INSEAD. Even more impressively, Singapore was ranked first for the innovation input sub-index, which comprises institutions (political/regulatory/business environments), human capital (education, R&D), infrastructure (especially ICT), market sophistication (credit, investment, trade and competition), and business sophistication (knowledge workers, innovation linkages, knowledge absorption). The country's forward-thinking policy-makers continue to focus on how to develop the nation to succeed in a future where high-technology manufacturing is rapidly shifting to newly emerging economies. One skill that many believe requires further development is creativity and innovation.[2] With her strong culture of entrepreneurship, it is no coincidence that many of the world's most innovative

Figure IV.2 From left: Quek Shu Yang and Xu Da Xiang.

ideas – such as Facebook and Google – came from the USA. Singapore is hoping to follow a similar path and has embraced the Silicon Valley model of creating a vibrant ecosystem for nurturing entrepreneurs and new venture-creation activities. A plethora of grants, incubation support, and research funding to grow this entrepreneurship ecosystem has been unleashed since 2009. One local firm that has benefited from new government initiatives is Gozo.

How it all began

For young Singaporeans, attending the National University of Singapore's Overseas Colleges (NOC) is an ideal way to see the world and learn the ins and outs of the start-up industry. The NOC program offers qualified students a minor in technopreneurship if they complete the dual requirements of a one-year internship with a start-up in one of the overseas locations along side taking one mandatory course 'new venture creation' at the partner university. Students can apply for one of seven locations, including Beijing, Philadelphia, Shanghai, and Tel Aviv. By far, the most popular location is Silicon Valley with the partner school being Stanford University, the US hotbed of entrepreneurial successes. Students are connected with mentors and resources, and can tap into a well of established connections for internships. Shu Yang wanted to be a part of the scene. His family did not have a history of entrepreneurship, but he was drawn to give back to his country and figured entrepreneurship was the best way to capitalize on his computer science degree. Practical and reserved, his rationality made him the perfect business partner for Daxiang. Confident and sometimes brash, Daxiang had poor primary school leaving examination (PSLE)[3] scores so he was unable to place in top secondary schools. He then started his tertiary education in Ngee Ann Polytechnic while waiting for an opportunity to enrol in NUS. He applied to enter NUS's hyper-competitive School of Business program and Nanyang Technological University (NTU)'s business school. His motivation to succeed was fueled by Starbucks CEO Howard Schultz's story, which spurred him to make a difference through entrepreneurship. Undeterred by the initial rejections, Daxiang wrote in an appeal by mobilizing supporters to write him letters of recommendation and won admission to both universities. Daxiang had heard about the NOC program from a friend's friend, and was immediately captivated by the idea. Daxiang developed his goal to enrol in NUS because he wanted to join the NOC program and go to Silicon Valley to run with technology entrepreneurs. The two vastly different Singaporeans first met as roommates after their acceptance to NOC Silicon Valley. Throughout their global sojourn, they supported each other through the joys and struggles of navigating the Valley ecosystem. It was during their stint abroad that the seeds of starting Gozo were planted, but they would not sprout for a few years.

Silicon Valley, a Californian suburb known for incubating some of the world's most valuable technology companies, continues to attract the world's brightest minds. Some of these talents flock to ubiquitous firms such as Facebook, Zynga, and Google, while others are keen to test their skills at one of many emerging start-ups. During their time in the Valley, NOC students sharpen their minds through classes at Stanford and build their professional clout through an internship. Hard working and reliable, Shu Yang interned at Valley-based Ultriva as a software engineer. At the web-based solutions company, he realized the impact technology would have on society and began thinking about how to make his mark. His housemate Daxiang had a very different experience. Bold, with a 'tell it like it is' attitude, he epitomizes many a young entrepreneur, but this trait caused him to clash with the head of the one-man start-up he was interning with. Fired after three months on the job, he was told by the NOC program office to get his act together and find another internship or be sent back to Singapore. He hustled to meet people, eventually outcompeting a number of Stanford students to land a coveted product management position at online retailer Zazzle. In a nutshell, working at Zazzle was fun, with energetic twentysomethings passionate about selling and marketing customizable wares. He drank in the energy, reveling in an atmosphere different from the school environments and rigid workplaces he had been used to. What he loved most about the company was their lack of hierarchy and constant dedication to improving – something he wanted to bring to his own company should he start one. Armed with knowledge, courage, and too many stories to count, they both returned home after one year to complete their studies at NUS before stepping away from entrepreneurship temporarily and onto public service and corporate warriors, respectively.

After graduation, Shu Yang had a solid stint at the Centre for Strategic Infocomm Technologies, a government-linked research and development unit. Despite his NOC experience, he did not think about pursuing entrepreneurship immediately. His first love was engineering and solving problems using technology. Besides, entrepreneurship did not run in his family and landing a good government job could be a stepping stone to more prestigious positions in the future. Or perhaps working for the government would give him the resources and stability to become an intrapreneur. Despite his carefully thought out plan, he was surprised that he soon became dissatisfied with the rigidity of his situation and yearned to create his ideal engineering culture.

Faced with the dilemma of wanting to hone his sales skills while lacking experience, Daxiang felt out his social network and interviewed for a sales position with Yahoo. Landing a position at one of the world's most valuable technology companies was no small feat, and he thrived in the competitive and innovative atmosphere. Before long, however, he again fell out with his superiors, disdaining workplace politics. A desire to bask in the prestige of his employer did little to tame his rebellious nature, and he soon admitted to himself that he was not suited for a corporate work environment. Sharing their frustrations with each other, they recognized that their differences made them the perfect team and agreed the time was ripe to pursue a venture of their own.

Keeping ears to the ground

During the summer of 2010, Shu Yang and Daxiang were gearing up to cut ties with their government/corporate safety nets and pursue their dream of start-up stardom. Nearing their late 20s, they figured it was now or never. Propelled by the confidence of having little to lose, they were certain they could grow a sustainable venture in the exploding group-buying deals environment. At that time, deal sites such as Groupon and AllDealsAsia were unerringly popular. But while consumers were attracted to the deep discounts, a number of participating merchants complained that the sites received too high a cut. Some even lost thousands of dollars by participating. The pair thought there was a profitable niche in offering similar discounts on mobile phones. Their big idea was to offer an auction-based model where merchants would bid to appear more prominently on their deals app. With Daxiang's extensive contacts built from his Yahoo stint, they were able to engage with multiple merchants on their idea. But through their conversations, they quickly realized that merchants were more intrigued with how the solution could be used to retain customers than acquire new ones.

What they needed was a way to turn new customers into repeat customers. Loyalty cards had long been a conventional method for incentivizing customers but were often misplaced before redemption. Businesses were also only able to incentivize customers who were already patrons of their stores. By combining location-based technology, customers could be alerted to participating merchants in their vicinity and redeem their coupons using QR code technology. With Shu Yang as technology architect, they quickly launched a mobile app called around!, signed on a number of merchants through their social network, and set out to test their idea. In August of 2011, the first QR code was redeemed. As far as they knew, they were the only company doing anything of the sort in Singapore. Excited as those who believe they are sitting on a golden egg always are, they were shocked to hear the news that Google had acquired a California-based start-up called Punchd who had a very similar value proposition. The company was unknown to around! but the founders knew that, as Singaporeans love reading blogs like TechCrunch, news would quickly spread and sprout copycats across the island.

Gozo time

The pending influx of copycats kicked the two young founders' adrenaline into high gear. Having been participants in the Valley's ultra-fast start-up ecosystem, they believed in their ability to knock out any competition in their home turf, and they officially quit their jobs in March of 2011. As young, educated twentysomethings, what did they have to lose? They considered the opportunity cost for 'sky is the limit' potential of the mobile apps field. Having benefited greatly from NOC, they decided to pitch to the NEC hoping to have their idea validated. Shu Yang shares, 'NUS gave us this energy. We pitched to see if someone could validate our idea and not only did they do that, but they also gave us money. We were amazed'. The pair had considered bootstrapping but the offer to participate in the incubation route was too good to turn down. The NEC invested S$50,000 in the business in return for a 5 per cent option. They were also given the equivalent of a five-star treatment for entering an incubator – an office on campus, free internet connection, access to consulting advice and development workshops, and assigned a business mentor. Introductions to angel investors and seed-fund investors were also organized.

Members of the new generation

In the short two years between when they left to Silicon Valley and now, NUS's start-up culture had changed dramatically. Long regarded as Singapore's top university, the campus is recognized for the quality of education it provides. Before heading to the Valley, Shu Yang had been studying at the School of Computing (SOC), but despite learning from renowned professors and working in high-tech labs, he was dismayed at the lack of resources for aspiring entrepreneurs. Facebook and Google were changing how people all over the world thought about technology, but there was little encouragement for local students to aspire for the same. To him, this seemed a misjudgment of where the industry was going. His time in the Valley further confirmed his instinct, showing him the groundbreaking impact students of technology could have.

Determined to share his fresh insight with peers, he returned to Singapore after his years aboard to be pleasantly surprised. The SOC had begun to make changes, shaking off the cobwebs and thinking outside of the box, emphasizing technology's potential and encouraging students to work on innovative projects. He got to know Professor Ben Leong who at that point had galvanized groups of students to create a visitor sign-in management system for a local welfare organization through the Computing for Voluntary Welfare Organisations movement, and grew motivated to find a niche where he could make an impact on society. Singapore is exceedingly adept at responding to the needs of the times and the environment was adapting to the growing realization that an educated workforce without visionaries was detrimental to the country's future success. Concurrently, Singapore's NRF was also creating a number of initiatives to fund seed-stage technology start-ups, targeting young technopreneurs exactly like Shu Yang and Daxiang. One of the key initiatives is a technology incubation program[4] which serves as a catalyst for seed-round investments into young technology ventures. They were in the right place at the right time to build their start-up on home ground.

Attracting true believers

A focus on supporting entrepreneurship began to spread across the country, and around! found many businesses were amenable to fresh ideas. Spurred by a supportive ecosystem, Shu Yang and Daxiang focused on developing their loyalty rewards idea but entered a frustrating period where they were unable to sign on merchants.

Eventually they realized they had a chicken and egg problem. As a young start-up, the ability to monetize their business is the key to sustainability, so they wanted merchants to pay. What they did not understand was that the best way to attract merchants was by offering them a free trial, and the best way to attract users was to have a lengthy roster of merchants. This initial misstep took around! longer than anticipated to gain momentum and, by late 2011, they had a number of competitors such as Perx, Squiryl, Pointpal, and Hachicode. Fierce competition in the Little Red Dot[5] made it hard to monetize, and they needed to attract more funding to expand operations.

When around! entered the NUS Enterprise Incubator, they had been assigned an advisor. The mentor was a marketing heavyweight that not only had clout in the services industry but genuinely believed in their idea. Understanding their need for investors, she brought her investment partners to check out their idea. At this point, dialogue was informal and they were not presenting formal pitches or explaining their business plan. To their ecstatic surprise, this was enough for them to receive their first term sheet that November from the investment partnership of the advisor. Although it was for a low six-digit investment amount, the injection would be sufficient for Daxiang and Shu Yang to continue proving their idea and fight off the competing copycats.

Every start-up dreams of the day where their hard work is validated by those willing to invest financially in their idea, but terms outlined in the term sheet was disappointing. Their investors wanted more decision-making power, asking for a greater number of board seats and financial control over fairly minor amount of expenditures. Against the advice of other mentors whom they consulted, they accepted the terms because they needed an influx of cash since they had only enough money in the bank to sustain themselves for another five months. As soon as they signed the term sheet, the winds of change blew.

around!

around! is a mobile app for users to 'tap' points or stamps from merchants who are signed onto the around! platform. Merchants would have a QR code at their location, typically at the POS terminal, and when the customer has paid for their purchases, they would offer the QR code for users to scan and earn points for their purchases.

Life in the fast lane

Their assigned mentor stayed out of the term sheet negotiation process as there was a conflict of interest – she was assigned by NUS who had an option. After Gozo agreed to the term sheet, she continued to resume her mentorship role, albeit a more withdrawn one so they would not receive differing signals from herself and the partner from the firm who was assigned to manage their situation. This allowed the firm to retain one point of contact to represent the interests of the collective.

Unfortunately, the lack of a check to balance their decisions left the around! team feeling lost when the investors began more forcefully to exert their influence. To turn the company around and achieve cash flow, the investors believed weekly meetings and constant checking in was the best method to keep the team on schedule. Shu Yang and Daxiang found the experience demoralizing, feeling themselves downgraded to mere employees. But since the shareholder agreement cementing the terms had yet to be signed, the support staff at the NUS Enterprise incubator advised them to pitch to iJAM Reload[6] for a potential additional source of funding. The caveat was that they would be ineligible for the grant once the investment money touched their bank account. So they chose to defer closing the investment and went after iJam. Preparing their pitch, they won the money through the NUS 'green lane'[7] in March 2012. They were ecstatic – winning iJam meant an immediate $50,000 grant, to be followed by another $100,000 if they secured another $100,000 angel investment. The grant money came in at the ideal time because they had been operating at a high burn rate. Understanding that the investment money was all but guaranteed, plus the additional booster of the iJam grant, they took their investors' advice to 'just spend' and be aggressive in their marketing efforts.

Concurrently to the founders' iJam pitch and grant award, the investors who issued the term sheet were accepted as a TIS incubator, which meant the investors could now invest a larger quantum by leveraging NRF's matching investment.

By now, Gozo had sat on the signed term sheet for over four months and still had not closed the investment. Because of the various choices made by the founders, the company was now running short on cash. Daxiang reflects:

> We took on unforeseen debt because we were told to spend money. Because of that, unfortunately we started to fall behind on payroll in February

> and March, and we had to borrow money in order to break even. The problem was that the investors themselves were beginners, so that made things very tough. Tensions between us and the investors began to rise.

He wishes they had then known the unofficial first rule of entrepreneurship is 'Do not run out of money'. Despite escalating tensions, Daxiang decided to take a break from all the drama to bring his parents on a long-awaited trip to the USA to visit his elder brother and he asked Shu Yang to push for closing on the investment. As a native Singaporean, he was adapting to the frigid weather while walking around Niagara Falls when he took a devastating conference call from his investors. Over the phone, they informed him that they were having second thoughts about signing the shareholder agreement! Because they were still interested in around!'s technology, they presented him with three options. The first was to merge with another company they were pitched to and interested to invest in, and that was in a complementary space as Gozo. If Gozo was willing to merge, the new combined company would gain the full $589,000 investment via the TIS route. The second option was to allow the investors to bring in a new CEO and his management team, and the two co-founders be relegated to employees. The final option was to part ways.

At this point, Gozo had signed on about 50 merchants and 20,000 users. While their drama was unfolding, their competitors were expanding at a higher rate. Perx, the leading competitor, had raised substantially more cash and had acquired over 100 merchants and 50,000 users despite being a follower to Gozo by four months. It seemed that Perx had the momentum to win the loyalty card on mobile app space. Perx, however, was comprised of non-technical founders. They had outsourced their development to an offshore development house in eastern Europe. The other two competitors were also struggling to win a toehold in what was now a fiercely contested space.

Daxiang and Shu Yang have imbued within Gozo a very close-knit team culture, being the ones who hire every single team member. Gozo also has a very strong engineering work culture in solving problems via technical solutions. Most important, Daxiang and Shi Yang are proud of their fledging start-up's corporate culture. Every team member works together like brothers in a family. Ironically, it is this close and friendship-centric culture which has made the investors wary of Gozo's execution ability. Potential investors felt that Da Xiang and Shu Yang were idealistic and inexperienced, which meant that they could not manage the team objectively, to the detriment to Gozo's execution and performance in the very competitive race.

To be or not to be?

There are no certainties in the start-up scene. Brilliant ideas are a dime a dozen; execution is what sets winners apart from losers. Often, decisions are forks in the road ready to steer an idea onto a different path. Investors and advisors can help entrepreneurs with decision-making by sharing resources and wisdom, but entrepreneurs must make the final decision. No one knows a budding company better than the founder who has conceptualized it and raised it from inception. The around! team currently has three difficult options. With a run rate of two months left, a staff of nine employees, and an ecosystem having difficulty monetizing. *Which of the three options should they take?*

Notes

Preface

1 The Institute of Systems Science was the premier IT research lab; it has since been split into two components. The present-day ISS is a post-graduate IT education institution. The technology research component has gone through successive name changes: Kent Ridge Digital Labs (KRDL), Laboratory for IT (LIT), and finally renamed to I2R, one of the 13 research labs under A*STAR.
2 Now the Agency for Science, Technology and Research (A*STAR), this is a government organization aimed at developing world-class scientific research and talent in Singapore to enhance the country's economic competitiveness. See www.a-star.edu.sg/ for details.
3 Bud Fox is a character in the movie *Wall Street*. A scene in the movie has the character having to sell everything to get out of a bad decision.
4 *Jianghu* – a Chinese term with martial arts origin to mean the ungoverned, sometimes chaotic community of practitioners. See: http://en.wikipedia.org/wiki/Jianghu for details.
5 See Chesbrough, 2005; Gompers *et al.,* 2005.
6 Disclosure: the main author is an angel investor in the ventures described in both teaching cases.
7 Effectuation – In broad strokes, the theory of effectuation (Sarasvathy, 2008) suggests that entrepreneurs use two different logic approaches when taking key actions in the creation of a new venture. Effectuation is means-based logic where entrepreneurs take actions based on what they know, whom they know, and who they are. This logic is in contrast to causation logic where entrepreneurs take actions based on some prediction on market trends and needs.

1 Star+Globe Technologies and Virginia Cha

1 *Wuxia*: Martial arts adventure stories were popular with youths in my day. My personal favorite writers were Louis Cha (no relation to the author) and Gu Long. See http://en.wikipedia.org/wiki/Wuxia.
2 *Youxia*: Protagonists in *wuxia* novels – usually depicted as wandering heroes who would go from place to place to fight evil and do good deeds. See http://en.wikipedia.org/wiki/Youxia.
3 Chinese for rich husbands' wives.
4 COBOL: Common Business Oriented Language, a high-level computer language in use for mainframe systems in the 1960s and 1970s. For the computer buffs, there were two variants – COBOL68 and COBOL72.
5 Burroughs: The 'B' in BUNCH companies – Burroughs, Univac, NCR, Control Data Corporation, and Honeywell. In the pre-mini-computer/Unix/personal computer days, IBM and the Bunch companies dominated the information technology industry.

6 Unisys – the name chosen to mean United Information Systems – after the merger of Burroughs and Sperry Univac.
7 HTML: hyperlinked text markup language.
8 Singalab was the first spin-out, a joint venture between ISS and IBM.

2 Pixelmetrix and Danny Wilson

1 *Archie's Girls Betty and Veronica* and the subsequent series *Betty and Veronica* are comic book series published by Archie Comics focusing on 'best friends and worst enemies' Betty Cooper and Veronica Lodge, both girlfriends of Archie Andrews.
2 Singapore was ranked first in the world for ease of doing business. See http://www.doingbusiness.org/rankings for details.

3 Sinetics Associates and Danny Chng

1 NCR: Previously known as National Cash Register, one of the original BUNCH companies.
2 Singapore was ranked first 1 in the world for best shopping avenue. See www.cnngo.com/explorations/shop/mystery-shoppers-rank-worlds-best-shopping-avenues-060,108 for details.
3 See www.secinfo.com/dsvRs.659p.htm for financials of Tokheim filed to SEC.

5 PlaNET and Ronnie Wee

1 National Service: Singapore has a mandatory two-year military service for all males at age 18. See http://en.wikipedia.org/wiki/Conscription_in_Singapore for details.

6 System Access and Leslie Loh

1 Malaysian currency (approximately 3.07 ringgit to US$1).
2 Author's note: It is legal. Leslie Loh is not committing a crime.
3 RDBMS: Relational database management system.

Part III Coaching the young

1 See GUESSS 2011 Survey, published by NEC. The survey showed that the proportion of intentional founders is higher in Singapore (37.5 per cent), as compared to OECD countries (36.3 per cent) but a lower proportion of active founders (1.1 per cent in Singapore versus 2.5 per cent in OECD countries). This seems to be an opportunity to convert intentions to actions for Singapore.
2 OECD countries in the GUESSS 2011 survey are Austria, Belgium, Finland, France, Germany, Ireland, Japan, Luxembourg, Netherlands, Switzerland, and UK.
3 SPRING is Singapore's agency to promote small and medium enterprises and entrepreneurship development: see www.spring.gov.sg.
4 See http://app.mof.gov.sg/data/cmsresource/ESC%20Report/Subcommittee%20on%20Growing%20Knowledge%20Capital.pdf for details.
5 For a listing of the various technology funding schemes under NRF's NFIE, refer to: www.nrf.gov.sg/nrf/uploadedFiles/TIS%20Press%20Release_FINAL%20(14%20Oct%2011)(1).pdf. For a complete description of NRF, see National Research Foundation of Singapore: www.nrf.gov.sg.
6 The author (Virginia Cha) has since angel-invested in Paywhere and is a shareholder.
7 SingTel Innov8 (Innov8) is a wholly-owned subsidiary of the SingTel Group. Innov8 is setup as a corporate venture capital fund with an initial fund size of S$200 million. See http://innov8.singtel.com.

8 The six selected fund managers are: BioVeda Capital II, Nanostart Asia, Raffles Venture Partners, Tamarix Capital, Upstream-Expara, and Walden International. See www.nrf.gov.sg/nrf/uploadedFiles/News_and_Events/ESVF%20Press%20Release.pdf.
9 In May 2012, HungryGoWhere was acquired by Singtel for S$12 million. See www.channelnewsasia.com/stories/singaporebusinessnews/view/12,02946/1/.html.
10 Brandology was acquired in 2011 by marketing company Media Monitors. See www.mediamonitors.com.sg/about-us/media-releases/media-monitors-acquires-social-media-monitoring-company-brandtology.
11 Tan Ying Lan, formerly with Singapore NRF, in an interview with the author, June 2012.
12 2009 TIS incubators: Clearbridge Accelerator, I2G Tech Accelerator, Neoteny Labs, Plug & Play, Social Slingshot, Small World Group, Stream Global.
13 2012 TIS incubators: Get2Volume, Golden Gate Venture Investments, IncuVest, Jungle Ventures, Red Dot Ventures, The Biofactory, The Network Fund (TNF), WaveMaker Labs. See www.nrf.gov.sg/nrf/uploadedFiles/20120309%20TIS%20Press%20Release%20(FINAL).pdf.
14 iJAM Reload is one of the many start-up funding schemes supported by Singapore NRF. iJAM is a grant-funding scheme administered by the Media Development Authority of Singapore to fund innovative ideas in the new media sector.
15 See www.temasek.com.sg/Documents/userfiles/files/Temasek%20media%20statement%20on%20CIP%20and%20CIP%20Details.pdf.

Conclusion

1 Part IV: Teaching case study 1.

Teaching case study 1

1 In 2012, Singapore ranked as the third richest country according to Forbes (see www.forbes.com/sites/bethgreenfield/2012/02/22/the-worlds-richest-countries/) and ranked as first in ease of doing business (see www.doingbusiness.org/~/media/fpdkm/doing%20business/documents/annual-reports/english/db12-fullreport.pdf).
2 According to a 2012 published report, the bottom five values in importance to Singaporeans are: curiosity, appreciation of beauty and excellence, creativity, spirituality, and courage (see www.straitstimes.com/BreakingNews/Singapore/Story/STIStory_773246.html).
3 Full disclosure: The author is the mentor and angel investor.
4 SPRING (www.spring.gov.sg) is Singapore's agency for promotion of small and medium enterprises.
5 National Research Foundation (www.nrf.gov.sg) has a technology incubation scheme to support incubators in funding technology-based startups in Singapore.

Teaching case study 2

1 In 2012, Singapore ranked as the third richest country according to Forbes (see www.forbes.com/sites/bethgreenfield/2012/02/22/the-worlds-richest-countries/) and ranked as first in ease of doing business (see www.doingbusiness.org/~/media/fpdkm/doing%20business/documents/annual-reports/english/db12-fullreport.pdf). INSEAD's 2012 Global Innovation Index ranked Singapore as first in innovation.
2 According to a 2012 published report, the bottom five values in importance to Singaporeans are: curiosity, appreciation of beauty and excellence, creativity, spirituality, and courage (see www.straitstimes.com/BreakingNews/Singapore/Story/STIStory_773246.html).

3 PSLE: The first of many hurdles Singaporean students must overcome to enter good schools. A poor PSLE score often places the student in a less prestigious secondary (equivalent to junior high/high school) with a lower chance of scoring high on the A-levels, the equivalent of high school leaving exam. A low A-level score would diminish the opportunity to enter the fiercely competitive local universities, with NUS being ranked top among the four local universities.

4 The TIS, inspired by Israel's Technological Incubator Program, seeks to address the gap of seed-round investments and build the entrepreneurial ecosystem by offering quality mentoring, funding, and space incubation from a number of approved incubators. The scheme de-risks early-stage investments by offering co-investment from NRF to match the investment from qualified TIS incubators. The qualified TIS incubator contributes expertise in selecting appropriate start-ups and is able to invest up to 15 per cent, or $89,000, while NRF acts as a joint investor contributing the remaining 85 per cent up to a maximum investment of $500,000. The scheme is structured as a convertible bond with the incentive to allow the incubator to buy out the NRF's $500,000 bond at the original value with a nominal interest in three years. The objective of this scheme is to encourage seed-stage investments and has the net effect of turning qualified angel investors into scalable incubators to fuel the entrepreneur ecosystem. The scheme was launched in 2009 and has 15 incubators selected as of 2012.

5 Little Red Dot is often a local self-deprecating term to refer to Singapore, a small red spot on the global map.

6 iJAM Reload is one of the many start-up funding schemes supported by Singapore NRF. iJAM is a grant funding scheme administered by the MDA of Singapore to fund innovative ideas in the new media sector.

7 Due to large number of grant proposals, some entities such as NUS incubation have the ability to pre-quality and send the selected candidates to the front of the grant review committee.

References

Adnan, H. (2002) 'VCs and entrepreneurs urged to narrow info gap', *The Star* www.ispringcapital.com/News762002.htm (accessed August 2012).

Baker, T. and Nelson, R.E. (2005) 'Creating something from nothing: Resource construction through entrepreneurial bricolage', *Administrative Science Quarterly* 50: 329–66.

Chesbrough, H. (2005) *Open Innovation: The New Imperative for Creating and Profiting from Technology*, Harvard Business School Press.

Gompers, P., Lerner, J., and Scharfstein, D. (2005) 'Entrepreneurial spawning: Public corporations and the genesis of new ventures, 1986 to 1999', *The Journal of Finance* 60 (2): 577–614.

Read, S., Dew, N., Sarasvathy, S.D., Song, M., and Wiltbank, R. (2009) 'Marketing under uncertainty: The logic of an effectual approach', *Journal of Marketing* 73 (3): 1–18.

Sarasvathy, S.D. (2008) *Effectuation: Elements of Entrepreneurial Expertise*, Northampton, MA: Edward Elgar Publishing.

Sarasvathy, S.D. and Dew, N. (2005) 'Entrepreneurial logics for a technology of foolishness', *Scandinavian Journal of Management* 21 (4): 385–406.

Index

For Product Safety Concerns and Information please contact our EU
representative GPSR@taylorandfrancis.com
Taylor & Francis Verlag GmbH, Kaufingerstraße 24, 80331 München, Germany

www.ingramcontent.com/pod-product-compliance
Ingram Content Group UK Ltd.
Pitfield, Milton Keynes, MK11 3LW, UK
UKHW011019180926
475701UK00004B/840

* 9 7 8 1 1 3 8 9 1 0 3 7 9 *